OXFORD BIOLOGY PRIMERS

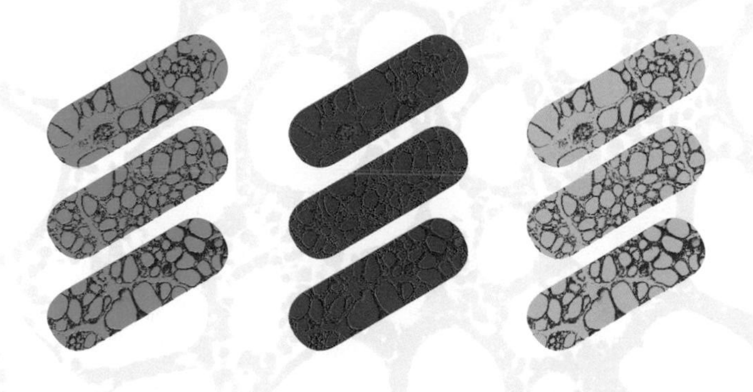

Discover more in the series at

www.oxfordtextbooks.co.uk/obp

Published in [illegible] the Royal Society of Biology

PLANT DISEASES AND BIOSECURITY

OXFORD BIOLOGY PRIMERS

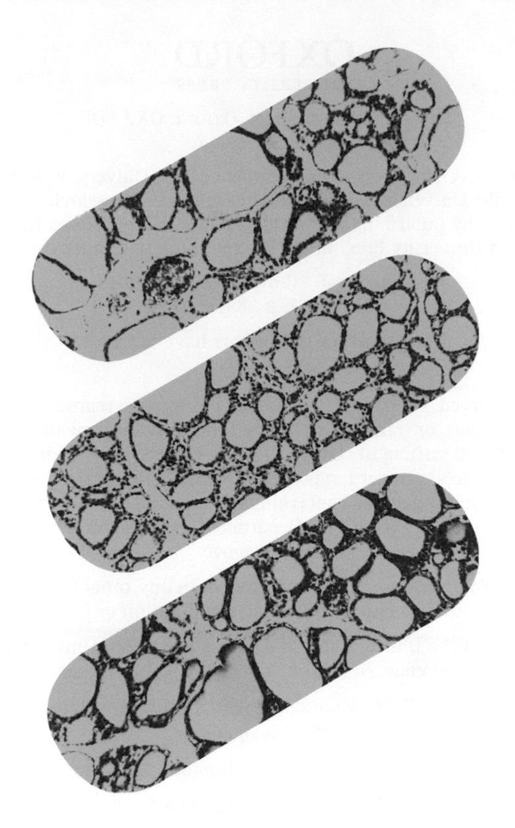

PLANT DISEASES AND BIOSECURITY

Paul Beales, Animal and Plant Health Agency
John Elphinstone, Fera Science Ltd
Adrian Fox, Fera Science Ltd
Charles Lane, Fera Science Ltd
Derek McCann, Animal and Plant Health Agency
Julian Little, Bayer

Edited by Ann Fullick

OXFORD
UNIVERSITY PRESS

Great Clarendon Street, Oxford, OX2 6DP,
United Kingdom

Oxford University Press is a department of the University of Oxford.
It furthers the University's objective of excellence in research, scholarship,
and education by publishing worldwide. Oxford is a registered trade mark of
Oxford University Press in the UK and in certain other countries

Impression: 1

Published in the United States of America by Oxford University Press
198 Madison Avenue, New York, NY 10016, United States of America

British Library Cataloguing in Publication Data
Data available

Library of Congress Control Number: 2019930582

ISBN 978–0–19–882772–6

Printed in Great Britain
by Bell & Bain Ltd., Glasgow

PREFACE

Welcome to the Oxford Biology Primers

There has never been a more exciting time to be a biologist. Not only do we understand more about the biological world than ever before, but we're using that understanding in ever more creative and valuable ways.

Our understanding of the way our genes work is being used to explore new ways to treat disease; our understanding of ecosystems is being used to explore more effective ways to protect the diversity of life on Earth; our understanding of plant science is being used to explore more sustainable ways to feed a growing human population.

The repeated use of the word 'explore' here is no accident. The study of biology is, at heart, an exploration. We have written the Oxford Biology Primers to encourage you to explore biology for yourself—to find out more about what scientists at the cutting edge of the subject are researching, and the biological problems they're trying to solve.

Throughout the series, we use a range of features to help you see topics from different perspectives.

Scientific approach panels help you understand a little more about 'how we know what we know'—that is, the research that has been carried out to reveal our current understanding of the science described in the text, and the methods and approaches scientists have used when carrying out that research.

Case studies explore how a particular concept is relevant to our everyday life, or provide an intimate picture of one aspect of the science described.

The bigger picture panels help you think about some of the issues and challenges associated with the topic under discussion—for example, ethical considerations, or wider impacts on society.

More than anything, however, we hope this series will reveal to you, its readers, that biology is awe-inspiring, both in its variety and its intricacy, and will drive you forward to explore the subject further for yourself.

ABOUT THE AUTHORS

Animal and Plant Health Agency (APHA)

The Animal and Plant Health Agency (APHA) is an executive agency of the Department for Environment, Food & Rural Affairs (Defra), and also works on behalf of the Scottish Government and Welsh Government to safeguard animal and plant health for the benefit of people, the environment, and the economy. The Plant Health and Seeds Inspectorate (PHSI) is part of the APHA and implements and enforces plant health policy. If you want to grow, import, export, or move certain plants or plant material, you will need to use the services provided by the PHSI.

Dr Paul A. Beales BSc (Hons), PhD, FRSB, SPHP (*Author of Chapters 1 & 2*)

Dr Paul Beales grew up in the countryside and had a fascination for biology from an early age. He concentrated on the sciences for his A levels, and studied biology in his first degree at Bristol, where he became fascinated with plant diseases. He then studied a new disease of rhododendrons for his PhD, and immediately afterwards Paul joined a government laboratory as a mycology (fungal) diagnostician. Here he had opportunities to design, develop, and use some of the latest techniques to identify plant pathogens (a kind of forensics for plants), to train plant pathologists around the world, to discover and publish brand new diseases, and communicate fungal plant diseases through journals, radio, and TV appearances. After 16 years in the lab, Paul moved to work with the UK's PHSI—the front line force that protects our environment through inspection services. Here he plays a big part in raising awareness of plant pests and diseases to everyone from school children to landowners, helping protect our environment for future generations.

Derek McCann BSc (Hons) FRSB SPHP (*Author of Chapter 6*)

Followings short stints in research and farming, and after studying Agriculture at Plymouth University, Derek has spent most of his career helping to hold the line to protect the plants of the UK so that we have food to eat, gardens to enjoy, and woods to walk in. He started as a front-line inspector in the North West of England before moving on to regional and national roles. As Principal Plant Health & Seeds Inspector, Derek currently leads on designing and delivering plant health surveys across the UK. He is also actively involved when notifiable pests and disease outbreaks are discovered in horticulture, agriculture, and the wider environment. Whilst we all have a role to play in plant biosecurity, Derek particularly appreciates the trust and responsibility the APHA are given as regulators working for the public good.

Fera Science Ltd

Fera Science Ltd is a national and international centre of excellence for interdisciplinary investigation and problem solving across plant and bee health, crop protection, sustainable agriculture, food and feed quality, and chemical safety in the environment. Fera Science's origins in delivering world-class science began over 100 years ago as the Institute for Plant Pathology services, and now it is a UK-based joint venture between Capita and Defra. Fera create and deliver integrated, innovative, and expert research services and products for our partners in crop protection, chemical, and animal health companies, as well as food producers and growers, manufacturers, distributors, and retailers. Fera also support and work closely with governments, academia, and leading research organizations.

Dr John Elphinstone, BSc (Hons) PhD (*Author of Chapter 3*)

Dr John Elphinstone has specialized in the detection, ecology, and control of plant pathogenic bacteria for over 30 years. He has a BSc in Plant Science from Newcastle University, and for his PhD in Dundee he investigated pathways of infection of seed potatoes by blackleg and soft-rot-causing bacteria, a topic that he is still actively involved with today. Following five years at the International Potato Centre in Peru, he spent a period at Rothamsted Research before moving to what is now Fera Science Ltd, where he has been based for the past 25 years. John has continued in his role as researcher in plant bacteriology and has regularly consulted with government and the EU on development and implementation of policy on monitoring and eradication of quarantine bacteria. John is a member of the European and Mediterranean Plant Protection Organization (EPPO) panel on diagnosis of plant pathogenic bacteria and a founding member of the European Association of Phytobacteriologists. He manages research projects on bacterial pathogens of all kinds of plants and mushrooms, as well as working on understanding the biological diversity of soils in relation to plant health and productivity.

Adrian Fox BSc (Hons) MRSB SPHP (*Author of Chapter 4*)

Adrian Fox is the Principal Plant Virologist at Fera Science Ltd. His work includes providing a plant virus diagnostic service and consultancy to the UK Government plant health service. He also conducts scientific research, which includes improving virus detection methods, plant virus discovery, and knowledge of virus transmission. Adrian leads a team which has been responsible for discovering and describing dozens of previously unknown plant viruses. He finds it really exciting to be able to show how these new viruses are spread and the range of plants they infect, with the potential to have a positive effect on plants of all descriptions. Although most of his work is focused on UK plant biosecurity, he also contributes to projects working on food security in Africa. Adrian really enjoys these projects, where applied plant pathology can make real and immediate improvements to people's lives and livelihoods. Adrian is a member of the International Committee on Plant Virus Epidemiology, and has published more than 30 scientific papers, new disease reports, and book chapters.

Dr Charles Lane BSc(Hons) PhD MRSB SPHP (*Author of Chapter 5*)

Dr Charles Lane studied Biology at the University of East Anglia for his first degree, and then completed a PhD in mushroom pathology at Sheffield University. Since then, for over 25 years Charles has worked in plant health and biosecurity, investigating the causes of new and emerging plant health diseases. For example, he was the first person to identify *Phytophthora ramorum*, the oomycete that causes sudden oak death, in the UK. In his role as a plant health consultant he has worked widely with government, non-governmental organizations (NGOs), industry, voluntary groups, and citizens in developing good biosecurity practice and raising awareness of plant health issues. He has also been leading the skills agenda for Defra and the Government Office of Science, developing a new programme of work to inspire the next generation of plant health scientists.

Bayer

Bayer is one of the world's leading life science companies and has interests in human health (including cancer and eye treatments and over-the-counter treatments such as aspirin), animal health (including flea treatments for pets), and crop science (helping farmers grow safe, high-quality, and affordable food, animal feed, and plant-based raw materials). With over 150 years' history, Bayer is Germany's biggest company. It has nearly 100 000 employees in 79 different countries, including the UK, where it employs nearly 1000 people. Bayer has two research farms near Cambridge, which combine the testing of new Bayer products with demonstrating how productive farming and the promotion of wildlife can go hand-in-hand.

Dr Tim Lacey BSc (Hons) PhD (*Contributor to Chapters 7 & 8*)

Having grown up on a farm, Tim has always been interested in agriculture—and particularly in plants. Initially, he followed this aspiration with a degree in Conservation Management, but then entered the world of arable and horticultural crops with a glasshouse research job for a major agrochemical company. Following this, Tim specialized further in horticultural crops with a PhD in irrigating vegetable crops before getting his hands dirty as an agronomist and triallist for several years. Ultimately, this led to his role looking after the specialist fruit, vegetable, and biological crop protection products at Bayer—a role which gives him great variety, is highly technical, and gives him the chance to talk to all stakeholders in the food chain—from growers to agronomists to supermarkets.

Dr Julian Little BSc (Hons) PhD FRSB
(Author of Chapters 7 & 8)

Dr Julian Little took a degree in Biochemistry at University of Swansea and then a PhD in how plants recognize (and do something about) the fact that they are being attacked by a fungus. He then started working as a plant scientist in Rhône-Poulenc, developing understanding of how weedkillers kill plants. Having worked in France, he returned to the UK as a project leader both in Rhône-Poulenc and then Aventis CropScience. When the latter was bought by Bayer, Julian switched to science communication, and he is currently the Head of Communication and Government Affairs for Bayer Crop Science. Julian has a lifelong fascination with plants and passionately believes that it is possible both to produce high-quality affordable food and to promote wildlife in our countryside. He was recently elected Fellow of the Royal Society of Biology.

Dr Kerry Maguire BSc (Hons) PhD
(Contributor to Chapters 7 & 8)

Dr Kerry Maguire has always been interested in plants. During her industrial placement year at Rothamsted Research (whilst studying for a degree in Plant Science at the University of London) she was introduced to the exciting world of soil remediation, where plants and microbes are used to help clean up pollutants. Then Kerry was introduced to plant pathology. Pathogenic microbes use stealth and disguise to infect the plant without being noticed until it is too late. Kerry embarked on her PhD, studying how a fungal pathogen infects wheat at a biochemical and cellular level, at Rothamsted Research UK and was awarded her Doctorate from the University of Nottingham, literaly becoming a plant doctor. Kerry stayed at Rothamsted before moving to NIAB TAG Cambridge, where she continued to work with cereal killers at a plant rather than a cellular level. She branched out to work with pea and bean pathogens before returning to her love of cereal pathogens. Now she works with scientists, growers, and agronomists to reduce the impact of cereal pathogens on crops.

Alice Turnbull BSc (Hons) MSc
(Contributor to Chapters 7 & 8)

Alice Turnbull is a Communications and Government Affairs Specialist at Bayer. Growing up in a family of scientists and engineers, Alice developed a passion for science communication and the need to connect scientists and others with environmental, food, and farming policy. She built on this passion by studying Sociology at the University of Nottingham, exploring the philosophy of science, and topics such as public acceptance of the science behind the climate change debate. After studying for a Masters in International Relations at Warwick University, Alice is delighted to be able to apply her research directly to the real world, working with a range of stakeholders every day on topics such as knowledge exchange, scientific innovation, and the power of plant science in shaping sustainable agricultural policy development.

CONTENTS

ABBREVIATIONS

cf.	compare
cm	centimetre
g/L	grams per litre
nm	nanometre
spp.	species

1 BASIC PRINCIPLES OF PLANT DISEASE

Dr Paul Beales

Introduction

Plants get sick! In fact all plants get sick, whether they are wild or cultivated. Some suffer only mildly with no visible symptoms; others succumb to more notable conditions such as leaf spots, leaf streaks and mosaics, stem cankers (depressions in the tissue), swellings, scabs, and galls, to name but a few. In severe instances infected plants may wilt and die.

Plant disease has been described since the dawn of recorded history. For example, the Bible and other early writings describe famines and social depravity caused by rusts, mildews, and smuts. Plant diseases affect the economy, environment, and social well-being of a country.

Records show the number of plant diseases that have entered and established viable populations in countries around the world to be increasing at an alarming rate. For example, a plant health risk register commissioned by the UK Government's Department for Environment, Food and Rural Affairs (Defra), published in 2014, highlights the risk of over 1000 plant pest and diseases. On average 88 new organisms are added annually. Over the past few decades the UK has seen an almost exponential increase in recorded plant diseases entering the country (Figure 1.1); some examples are given in Figure 1.2. This increase is primarily due to changes in farming practices, climate, and global trade; the latter results in plants moving around the world at a rapid rate.

During the times of the well-known Victorian plant hunters, ships were the only way of moving plants from one continent to another. If a symptomless but infected plant was put on a boat, it could be many weeks before it arrived

Figure 1.1 Increase in plant pest and disease over the past century

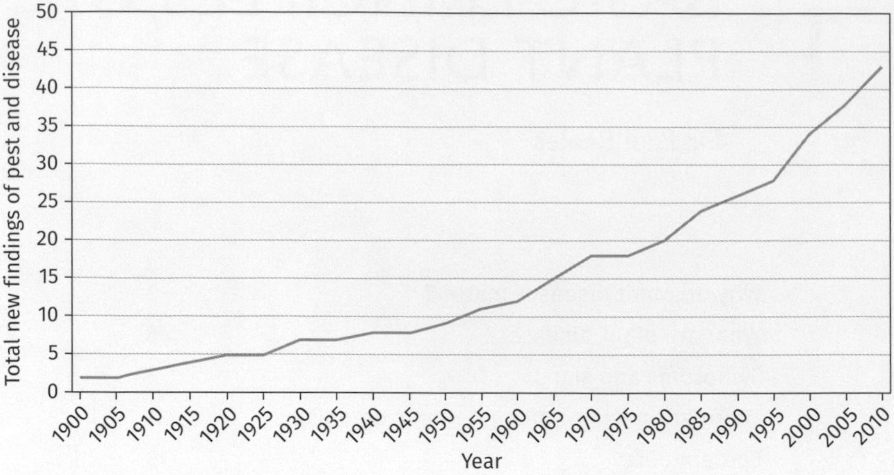

Total new findings of pest and disease (0–50) vs Year (1900–2010)

Figure 1.2 Some examples of plant diseases that have invaded and impacted UK plants in recent decades

1970
- Dutch Elm disease
  Ophiostoma novo-ulmi

1987
- Rhizomania
  Beet necrotic yellow vein virus

1992
- Potato brown rot
  Ralstonia solanacearum

2002
- Phytophthora ramorum dieback
  P. ramorum

2003
- Potato spindle tuber viroid
  PSTVd

2004
- Bleeding canker of horse chestnut
  Pseudomonas syringae

2006
- Phytophthora pseudosyringae
  P. pseudosyringae

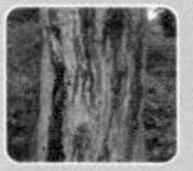

2010
- Acute oak decline
  Various (primarily bacterial)

2010
- Phytophthora lateralis
  P. lateralis

2012
Raspberry leaf blotch virus

2012
- Chestnut blight
  Cryphonectria parasitica

2012
- Chalara ash dieback
  Hymenoscyphus fraxineus

at its destination port, by which time any disease present would have been spotted (as the plant was dead!), and it would have been disposed of. With today's global transport mechanisms, however, plants can be growing on one side of the world one day, and in markets on the other side the next morning. It's not unknown to handle fruit that was growing on a bush 2000 miles away the previous day and find it is still warm! This highlights the need for good global plant biosecurity practices, both at the point of source and with border controls, as you will see in Chapter 8.

Why do plant diseases matter?

Before we get stuck into the details of plant diseases—what causes them, what they do, how they are spread, and how we can (or can't) stop them—it is worth taking the time to consider why they matter. It is easy to think that animals, and in particular human beings, are the most important organisms on Earth. But without plants and the resources they give us—from the food we eat and the oxygen we breathe to the clothes we wear and the homes we live in—the animal kingdom, including the human race, would be unable to survive.

Plant diseases affect many aspects of life, including the economy, environment, and social well-being of a country.

The economy

Economically, the impact of plant pathogens is evident in many ways. These include:

- the value of crop loss
- reduction in yield
- aesthetic losses
- cost of control measures
- diagnostics
- consultancy
- training
- monitoring
- outbreak management
- clearance (e.g. tree felling)
- loss of land for future cropping or social activities
- increased reliance on imports/restrictions of trade
- increased consumer price
- ecological costs.

A study carried out in Australia by Murray and Brennan in 2009 investigated the impact on the economy of 40 pathogens of barley over ten years. It was estimated that pathogens caused the average loss of $252 million per year, or 19.6% of the average annual value of the crop. Just three pathogens caused the greatest proportion of this loss—and this is when chemical and cultural management practices were applied. Without controls, it was estimated that each of the most significant plant pathogens would cause as much economic loss as all 40 pathogens combined. It is therefore essential to diagnose plant problems, identify the causal agent, and understand its epidemiology so that appropriate and effective management practices can be applied. All these aspects are covered in this book.

The environment

Plants are the foundation of the environment, supporting it through improving air quality and humidity levels, sustaining life to a myriad of fauna and flora, regulating the water cycle, and sustaining atmospheric oxygen levels. When plants or entire species are lost from an area due to a plant disease, the effect on the local environment can be devastating. For example, Dutch elm disease (*Ophiostoma novo-ulmi*) killed over 30 million trees in the UK in just a couple of decades (Figure 1.3). This resulted in a loss of habitat for over 60 species of insects, reduced food sources for birds, and increased soil erosion in some areas of the country.

Social well-being

Plants play a significant role in society and cultures throughout the world. Not only do they provide food, fuel, medicines, and shelter, but

Figure 1.3 An elm tree killed by Dutch elm disease, with the tunnels made by the beetles which carry the deadly fungus into the tree clearly visible

Anthony Short

they reduce stress: they make people happier, calmer, and better able to concentrate—and as a result, they also reduce crime rates. There is growing evidence that exposure to trees and other plants increases healing and recovery from surgery, reducing the length of time people need to stay in hospital. It is currently estimated that plant disease results in an average of about 5–10% loss in food production in developing countries (up to 40% in some)—and starvation has a major impact on social and physical well-being. Many businesses rely on plants for sports and social activities, and research in the USA has shown that by planting a few trees in a low-income area of Philadelphia, house prices were boosted by 2%.

The impact of plant disease can therefore be significant. During the outbreak of *Phytophthora ramorum* (sudden oak death) in California around the turn of the century, over 1 million tan-oak (*Notholithocarpus densiflorus*) trees were killed. Native Americans relied on this plant for food and business and the impact has been devastating to their traditional culture.

Further examples of the impact of plant disease on the environment, economy, and social well-being can be found throughout this book.

The rest of this chapter provides an overview of some of the basic principles of plant disease, what they are, how they are spread, what they do, how they are organized, how to identify them, and their importance throughout history. The rest of the book will take you deeper into all of these areas, and show you the importance of preventing and overcoming as many of these diseases as possible.

What are plant diseases?

When you think about the phrase 'plant disease', what does your mind conjure up? Terms such as dead, smelly, rotten, misshapen, and mouldy plants are common answers to this question. However, what most people describe are the symptoms or visual results and impact of disease on a plant. Plant disease is actually the result of a biotic or abiotic agent that has a harmful effect on a plant's normal growth or function. Note that, although the causes can be biotic or abiotic—living or non-living—the majority of this book will focus on biotic plant disease: disease that is a direct result of infectious microorganisms. Disorders such as drought stress, environmental damage (e.g. lightning, hail, wind), or disorders caused by the lack of certain vital nutrients or cultural damage due to poor plant management will be left out of this text. This book also does not, on the whole, include animal pests (e.g. insects), although they will be discussed as vectors of microorganisms.

Plant pathology, also termed phytopathology, is the study of plant disease caused by pathogens (including fungi, bacteria, viruses, and protista, as described in this book), their interaction with the environment, and host plants. It includes management practices designed to reduce or prevent the spread of disease.

Case study 1.1
The birth of plant pathology

If there is one plant disease almost everyone has heard of, it is potato blight. Potato blight is caused by *Phytophthora infestans*, which almost brought Ireland to its knees in the nineteenth century. Less well known is the story of *Plasmopara viticola*, and its impact on the way we understand and investigate plant diseases right through to the twenty-first century.

Around the same time potato blight was having such an impact on the population of Ireland, grape leaves in the Rhône valley, France were turning red and wilting. The grape crop shrivelled and became useless for wine making. The following year, the plants became stunted and died. The causal agent was determined as a root-infecting aphid (*Viteus vitifoliae*), commonly referred to as grape phylloxera. Wine-growing regions in North America also had grape phylloxera, but vines were showing resistance to the pest. The possibility of grafting North American rootstock onto French vines was proposed. Although initially reluctant to mix North American and French varieties, the wine industry in France was so severely impacted by grape phylloxera they actually had no choice and the rootstock was imported and grafting carried out. (Compare this with current arguments surrounding genetic modification —see Chapter 6.)

Unbeknown to the French, they inadvertently imported a microscopic stowaway on the American vines, the downy mildew organism *Plasmopara viticola*, which is biologically similar to *Phytophthora*. Downy mildew on grapes acts in a similar way to blight on potato. Initially, older leaves turn yellow and a white, velvety (or downy) appearance of the sporing structures (**sporangia**) forms on the lower leaf surface. As the disease progresses, the leaves wither and fall off.

The disease was known to occur in North America, and was actually encouraged, as mycologist William Farlow is quoted as saying, 'Our native vines have a luxurious growth of leaves, and the danger is that in our short summers the grapes will not be sufficiently exposed to the sun to ripen. The *Plasmopara* appears at just the right moment to shrivel up the leaves so that the direct rays of the sun may reach the grapes.' This, however, was under the drier climates of North America. Farlow also suggested that if the disease should occur in moister climes, the effects on grapevines would be much more devastating—and he was correct. Downy mildew began to affect the vineyards in France and soon spread across other wine grape growing regions of Europe, including Germany and Italy, killing the plants in its wake, and once again threatening Europe's valuable wine industry.

An inquisitive French botanist and mycologist called Pierre Millardet is believed to have been passing a vineyard that was affected by grape downy mildew one day. As a scientist, he was observant and noticed an area of plants

Figure A Shrivelled grapes and dying leaves caused by grape downy mildew. The blue Bordeaux mixture on the leaves of this grapevine will protect it from this disease.

Starover Sibiriak/Shutterstock.com

Denis Pogostin/Shutterstock.com

near the roadside that was unaffected by the mildew. Although there was no downy mildew present, he noted a mysterious bluish tinge to the leaves, which he did not recognize, so he went to ask the vineyard owner. The owner had been frustrated by people stealing his grapes, and so had mixed a concoction of chemicals to spray onto the leaves to deter the sticky fingers of the local grape lovers. It just so happened that amongst the mixture of chemicals were copper sulphate and lime (later named Bordeaux mixture after the place it was discovered). This is a potent mix of chemicals that, even today, will kill grape downy mildew (Figure A).

We mentioned earlier how downy mildew and potato blight were related. As a result, Bordeaux mixture also had a great effect as a killer of potato blight. However, this plant pathology story does not end here. Despite having a control for potato blight, a further severe outbreak occurred in Germany in 1916 in the heat of the First World War. It is believed to have contributed to as many as 700 000 deaths, and may have even been a factor in turning the tide of the stagnant First World War. This devastating epidemic occurred because copper normally reserved for the manufacturer of Bordeaux mixture to control potato blight and downy mildew on grapes was being used in the manufacture of shell casings and electric wire!

As a direct result of *Phytophthora infestans* and *Plasmopara viticola,* the study of plant pathogens, or plant pathology, was born.

Pause for thought

What steps could be carried out to prevent the introduction and spread of a new plant disease?

Symptoms and signs

The vast majority of infections by plant pathogens result at some point in an obvious outward symptom on a plant.

Symptoms may arise as a direct result of a pathogen damaging tissue, or may be indirectly produced by the plant itself—see Figure 1.7. The former is often seen with facultative pathogens. Such organisms produce enzymes that break down plant tissue on which they feed (e.g. plum brown rot caused by the fungus *Monilinia* secretes a range of plant cell wall and pectin-degrading enzymes). Indirect symptoms are normally the outward appearance of the host plant defending itself against the pathogen. The host plant may sacrifice its own plant cells surrounding the point of infection, in effect 'cutting out' the disease. This symptom is known as shot-holing and can often be seen on *Prunus* (cherry trees) infected with the fungus *Stigmina carpophila* (Figure 1.4).

Figure 1.4 Almond (*Prunus* spp.) showing shot-hole symptoms caused by *Stigmina carpophila*

Nigel Cattlin/Alamy Stock Photo

Internally, the host may form plugs within the xylem vessels (tyloses), which block pathogenic organisms from moving through the xylem and infecting other parts of the plant. However, with aggressive pathogens such as Dutch elm disease (*Ophiostoma nova-ulmi*), so many vessels are blocked with host-produced tyloses that the plant cannot get enough water to cells, resulting in the outward symptom of wilting and subsequent death.

Occasionally, a plant pathogen disrupts the normal metabolic pathways of its host, resulting in excessive or uncontrolled production of hormones such as auxin. The outward resulting symptom is cell hypertrophy—swellings, galls, and contortions (Figure 1.5).

Infection caused by obligate pathogens may result in indirect symptoms that are slow to appear and may only be visible as leaf yellowing. This is because the pathogen is stealing massive amounts of nutrients from the

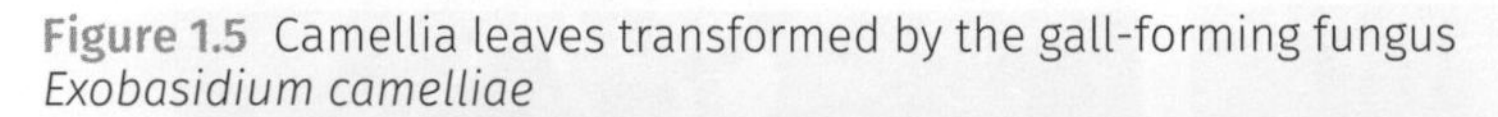

Figure 1.5 Camellia leaves transformed by the gall-forming fungus *Exobasidium camelliae*

GFK-Flora/Alamy Stock Photo

plant to use itself whilst keeping the plant alive. A good example is during a powdery mildew attack—look at young oak leaves in the late summer infected with the powdery mildew fungus (*Erysiphe alphitoides*); they remain green, but a copious amount of white fungal spores on the surface reveal that they are actually infected.

Warning! It is worth noting at this point that an obvious symptom such as wilting, dieback, or early leaf drop observed in a plant may be as a result of a root/lower stem infection, not where the symptoms are most obviously manifest. A good plant pathologist will take a holistic look at plants when examining symptoms.

To a plant pathologist, the signs of a disease are *not* the same as the symptoms caused by the infection. Signs of a disease are the visible presence of the organism itself. Symptoms are the effect of the infection on the plant. So the sign of oak powdery mildew mentioned in the previous paragraph is the white, powdery appearance to the leaf, which consists of fungal hyphae and spores—you can see this powdery appearance in Figure 1.6. The yellowing is a disease symptom.

If we look at roots, symptoms of a disease such as *Phytophthora cactorum* may be rots and decay, but the signs are the structures that produce reproductive spores called sporangia. In high numbers these may just be

Holger Kirk/Shutterstock.com

Krasowit/Shutterstock

Figure 1.6 The characteristic powdery appearance of oak leaves affected by oak powdery mildew is particularly noticeable next to healthy green leaves. A scanning electron microscope reveals the individual hyphae, sporing structures, and spores on the surface of the leaf.

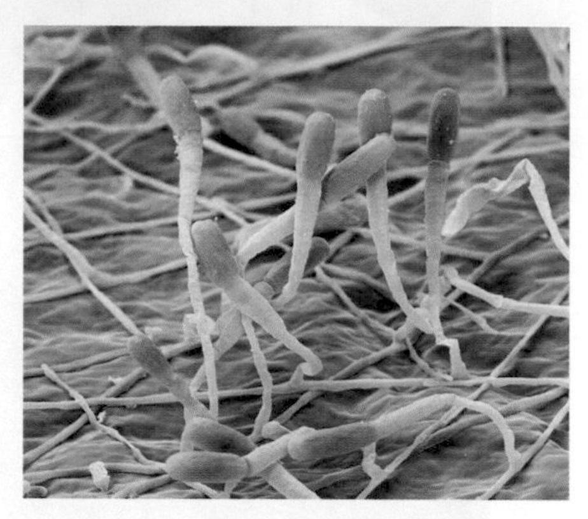

Eye of Science/Science Photo Library

visible with the naked eye, but usually some form of optical magnification is required.

Of all the groups of plant pathogens described in this book, signs of fungal diseases are usually the most obvious. There are, however, some diseases of bacteria which can be visible to the naked eye. For example, some trees infected with a bacterial pathogen release a thick, sticky mass of gum (termed gummosis) that consists entirely of bacterial cells. These ooze from infected trees and are easy to see. As viral infections occur at the cellular level, they are unseen, and hence do not produce visible signs, only symptoms.

A pathogenic organism may be present in some plants, but may not show any symptoms or visible signs, i.e. the plant remains healthy. This type of infection is termed quiescent. In some hosts, a change in the environmental conditions may 'wake' the pathogen resulting in disease, but in others the pathogen is like a stowaway, until a sap-sucking insect picks it up and transports it to a host plant that it can infect. For example, potato spindle tuber viroid is symptomless in some ornamental *Solanum* spp. If it is picked up by a sap-sucking insect and transported to pepper plants it causes mild leaf symptoms, but if it infects a tomato plant it causes severe symptoms, resulting in a disease known as tomato bunchy top.

Figure 1.7 Some of the common symptoms of plant diseases

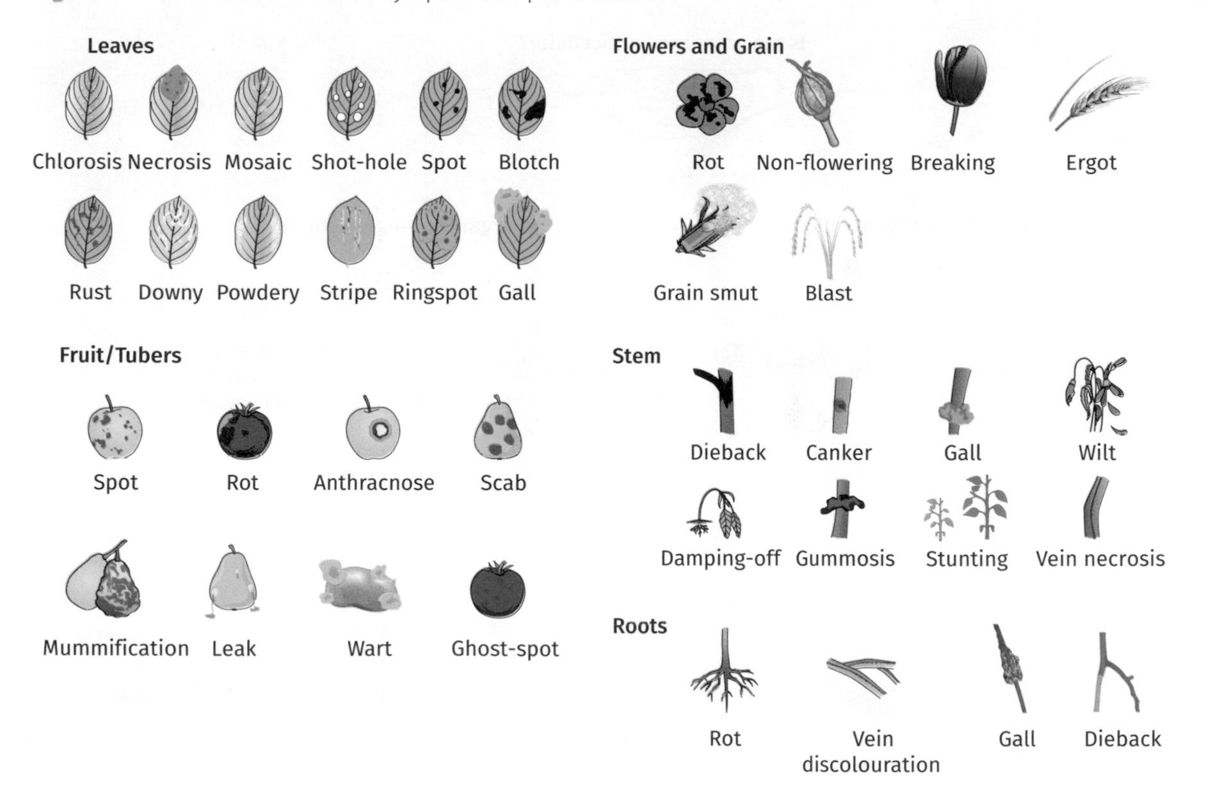

Common symptoms of plant diseases

Let's take a journey around a typical plant and identify some of the common symptoms of plant disease. Figure 1.7 shows some of the symptoms of disease which may be seen on different parts of a plant as a result of a pathogenic infection.

Major groups of plant pathogenic organisms

This book will study the following taxonomic groups: fungi, protista, bacteria, viruses, and viroids. The protista contain a group of fungal-like organisms classified as oomycetes that are responsible for many devastating plant diseases such as potato blight and grape downy mildew. Figure 1.8 shows you how each of these relates to other living organisms.

Plant pathogens can be divided broadly into two groups, biotrophs and necrotrophs. Biotrophs have specialized mechanisms to keep the host plant alive. They take the energy they need for growth and reproduction from living cells. Necrotrophs, on the other hand, derive all their energy for growth and reproduction from dead cells—either by killing them directly, or indirectly by feeding on cells killed by something else. There is one further group named hemibiotrophs that are somewhere in between the two—part

Figure 1.8 A basic key to separate major groups of plant pathogens

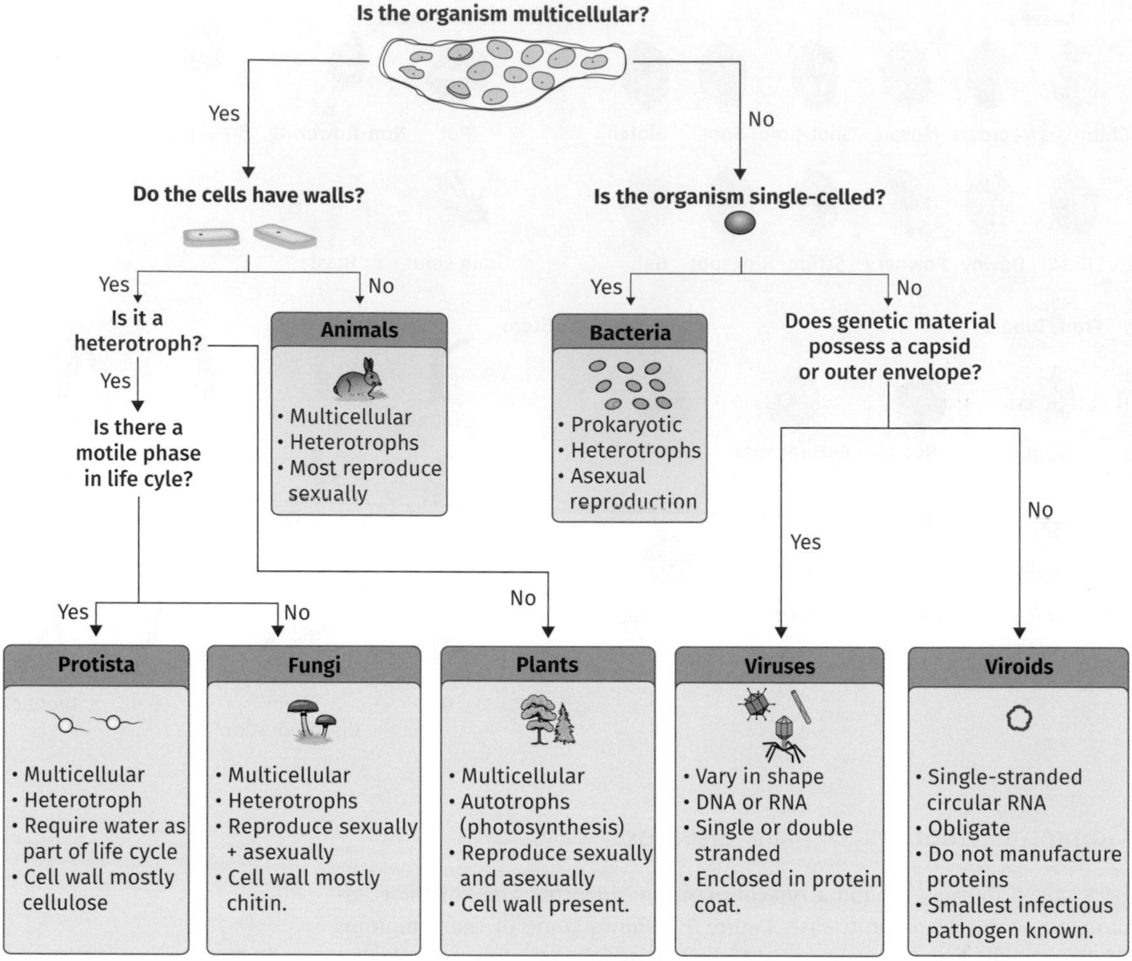

of their life cycle is biotrophic, but after some time they convert to necrotrophy. All viruses are biotrophs; bacteria can be biotrophs (e.g. *Ralstonia* and *Xanthomonas*), necrotrophs (e.g. *Pectobacterium*), or hemibiotrophs (e.g. *Pseudomonas syringae*); and fungi and protista can also be all three—biotrophs (e.g. powdery mildews and rusts), necrotrophs (e.g. grey mould (*Botrytis cinerea*)), and hemibiotrophs (e.g. potato late blight). All of these relationships involve a one-way flow of nutrients from the plant to the pathogen with no benefit to the plant (i.e. a parasitic interaction).

Classification, taxonomy, and nomenclature

The classification of plant diseases involves identifying similarities between organisms and subsequent arrangement into taxonomic groups, which can then be named accordingly (nomenclature). By grouping organisms in such a manner, it is easier to develop a better understanding of how each group functions. For example, if a new plant pathogen is discovered, understanding which broad taxonomic group it fits in will enable the researcher to have a basic understanding of its potential pathogenicity, epidemiological characteristics, and appropriate controls that might be effective against it.

Epidemiology

In many cases, a disease spotted and identified on a particular plant has, is, or will begin to spread, resulting in an epidemic. The study of epidemics is known as epidemiology and examines the complex nature of interrelating factors that enable a plant pathogen to establish a viable population in a given location. Figure 1.9 shows what is referred to as the *disease triangle*. It can be compared to the fire triangle and just as a fire requires an ignition source, oxygen, and a flammable substrate to establish, a plant epidemic requires the pathogenic organism, host, and a suitable environment. This representation could be extended to a disease tetrahedron, if human influence is included, as you can see in Figure 1.9.

The more information that can be gained through research into each of the areas in the disease triangle and tetrahedron for any particular plant health problem, the better the disease management that can be achieved.

When you are attempting to understand a plant disease more fully, some information will be readily available such as host, location, taxonomy, range of plants affected, etc. However, understanding some of the more cryptic factors such as how susceptible or resistant the host plant is over a period of time as it interacts with the pathogen may require molecular-level examination of factors such as plant defence mechanisms, e.g. the activation of effector genes.

To help get a better understanding of plant pathogen risk and biology, artificial experiments are carried out under controlled conditions. Factors such as temperature, humidity, rainfall, and light conditions can then be altered individually to help ascertain the conditions that would favour an epidemic.

A good understanding of the pathogen's life cycle, including how the disease is spread, is critical to understanding how an epidemic arises. As with the artificial experiments described above, some elements of the life cycle can be quickly established, but others, such as identifying where the pathogen survives adverse conditions—which may be in the host, the soil, water, plant debris, alternative host plants (this is common with many rust fungi), or insects (e.g. aphid-borne viruses)—can be much more difficult.

A life cycle is rarely fixed and can vary depending on the infected host and environmental conditions. It can be affected by the environment at both the macro (wide) scale and the micro or local scale—e.g. the microclimatic conditions surrounding the leaf. The addition of human influence complicates the epidemiological web further. For example, the application of pesticides may result in changes to the pathogen (e.g. it becomes resistant) and the movement of infected plants increases the risk of rate of spread.

Figure 1.9 The plant disease triangle. When humans become involved, it becomes the plant disease tetrahedron.

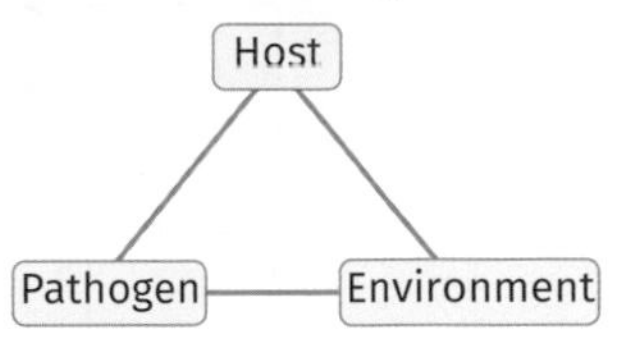

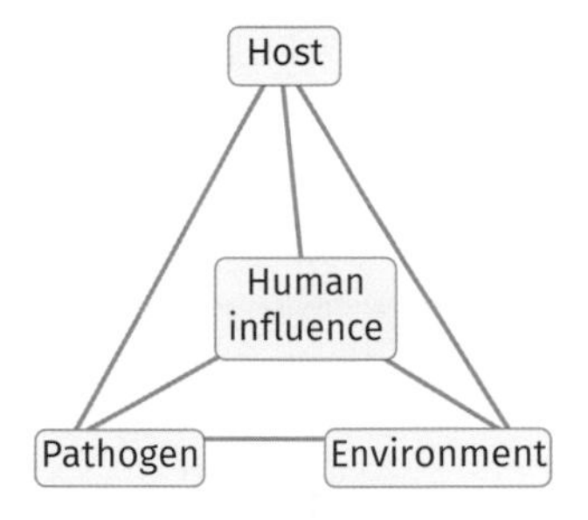

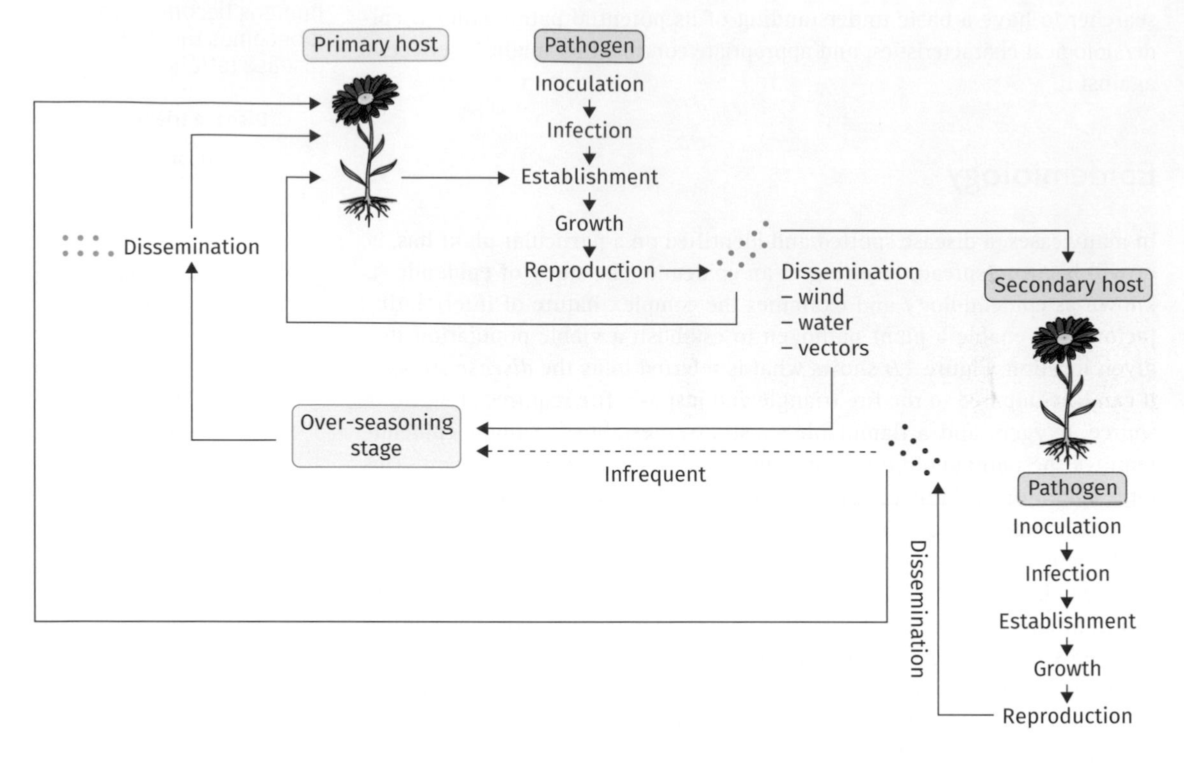

Figure 1.10 A simplified life cycle of a plant pathogen. Note: not all stages occur for every plant host/ pathogen interaction.

Generalized life cycle of a plant pathogen

We can describe the life cycle of a pathogen in a number of different ways. A common method is to follow the natural seasons (spring, summer, autumn, and winter) and identify what stage the pathogen is at each. Alternatively, you can just describe one complete pathogen growing cycle and include host interactions at relevant places. Figure 1.10 shows a generalized pathogen life cycle.

How plant diseases spread

Before we can try to control or eradicate plant diseases, we need to understand how the pathogens are spread from one host to another. Some methods of spread are common to all types of pathogens and others are more specific.

Local spread of pathogens

There are a number of ways in which plant pathogens can spread locally:

- **Soil-borne spread**—Some pathogens or their spores can survive long periods in bare soil or around the roots (**rhizospheres**) of crops and weeds. These plants may themselves be true hosts or they may simply

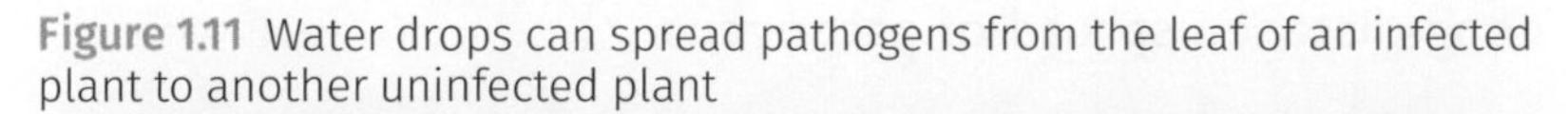
Figure 1.11 Water drops can spread pathogens from the leaf of an infected plant to another uninfected plant

lostbear/Shutterstock.com

supply nutrients as they die and decay to nurture the pathogens until the real host plant is planted again.

- **Spread in irrigation water**—Some pathogens infect plants which grow along river banks and the edges of lagoons. Pathogens entering the water can then be transported to the spores of host plants they come in contact with. Pathogens can also spread rapidly from diseased to healthy plants in hydroponic growing systems through the nutrient solution which is circulated around the crop.
- **Rain splash/wind dispersal**—Wet and windy weather is ideal for spreading disease. Rain (or overhead irrigation water) landing on infected plants picks up some types of pathogens and distributes them to nearby hosts (Figure 1.11). Wind and heavy rain may also damage plants, making it easier for pathogens to enter and infect plants through freshly formed wounds.
- **Insect transmission**—Some plant pathogens survive and multiply in the guts of certain insects. If one of these insects feeds on a healthy host plant, it can infect it with the pathogen. Alternatively, an insect can pick up a pathogen from an infected plant and spread it to the next healthy host it visits. Such disease-spreading insects are known as vectors.
- **Contaminated tools and machinery**—We should not forget the important role that people, and their modern ways of growing and managing crops and the natural environment, can have in spreading diseases. Machinery and tools, when used in infected crops, can be easily contaminated with pathogens, which can then be transmitted to the next healthy crop visited. Systemic infections are particularly easily transmitted during pruning activities, when the pathogens can pass from infected cutting blades directly to cut and exposed vascular tissues.

Long-distance spread of pathogens

Plant pathogens are often spread long distances by human intervention. People move infected seeds or other propagating material (young plants, cuttings, bulbs, tubers, corms, rhizomes, etc.) for growing in a new field, area, or country where a particular disease has not previously occurred (Figure 1.12). Since the pathogens are already surviving inside the infected host, once suitable conditions are present in the new environment (remember the disease triangle) a new outbreak of disease can begin—this is a Trojan horse of the plant pathology world! It is very difficult to intercept such infected consignments and prevent them from being planted. This is especially the case if the incidence of infection is low or where infections are latent and symptomless, so they may escape detection during inspections, sampling, and testing.

Figure 1.12 Both plants and seeds are transported all over the world—and they can take the pathogens that cause devastating diseases with them

photobar/Shutterstock.com

Peter Gudella/Shutterstock.com

A number of plant diseases are seed-transmitted. If the surface of the seed is contaminated, the risk of the disease spreading may be significantly reduced by, for example, disinfecting the seeds. However, some vascular pathogens can infect seeds internally, where they can remain protected and survive both disinfectant treatments and storage periods.

Identifying the causative agent of plant disease

Diagnosing what has made a particular plant sick parallels forensic science in humans and could be regarded as 'Crime Scene Investigation' for plants! As we have seen, a disease requires the perfect storm of host, pathogen, and environment to be successful. In order to carry out an accurate diagnosis, all of these aspects must be considered, investigated, and recorded. You will be looking at this in more detail later in this book—see Chapters 2, 3, 4, and 6—but here are some of the key factors:

1 Location

This is very important to help towards a diagnosis. Is the diseased plant isolated or surrounded by others in the same plant genus, or different genera (a knowledge of plant identification helps here)? An isolated plant will lead to the question: how did the disease get there? If there are surrounding plants, are they also infected? Many plant pathogens are host specific, others have a much wider host range. Volunteer plants growing in agricultural land can carry over disease from one season to the next. Is the plant in the shade or full sun? Fungi and bacteria in the main prefer darker, damper environments to survive. What is the plant's aspect? Is it in a dip or at the top of a hill? Low-lying areas are often less well drained than higher ground. Waterlogged soils can lead to higher instances of water mould infections such as *Phytophthora* or *Pythium*. These pathogens also survive in running water, so check whether there are any water courses in the location. What is the soil like? Compacted, overly dry, or wet soils can stress a plant, making it more susceptible to opportunistic pathogens (such as a number of *Pythium* spp., or *Erwinia* bacteria). Is the plant growing in good, nutrient-rich soil or not? Again, poor soil may not be providing sufficient nutrients to the plant. At this stage it would be worth establishing whether any pesticides have been applied to the field and, if so, which chemical and how often it was applied. For example, herbicides to kill weeds may have drifted onto the dying plant you are investigating, so no pathogen is involved at all.

2 History

If possible try and find out a little bit about the history of the site and the problem. For example, is this the first sighting of the problem? If it has been seen previously, how often, and is the problem getting worse or remaining stable? Some diseases such as stem cankers will naturally heal over time as the plant fights off the pathogen attack. Have there been any recent plant introductions? Diseases may not be evident on certain host species and survive as latent infections. When a suitable host and environment is present, a latent infection can develop or spread to more susceptible hosts. What was growing in the soil prior to planting the affected plant? Diseases such as clubroot (*Plasmodiophora brassicae*) can survive in a resting spore state in the soil for decades following an ancient infected brassica crop such as cabbages or broccoli.

3 Weather

This plays a vital role in disease establishment. Knowledge of key factors such as prolonged rain or drought, temperature fluctuations, humidity, prevailing winds, or even hail and thunder can aid diagnosis of disease. Disease prediction models are based on knowledge of prevailing weather conditions, and help with disease management. For example, the Hutton criteria state that a potato blight risk increases when there are two consecutive days with a minimum temperature of 10°C and at least six hours with a relative humidity $\geq 90\%$.

Many suspected diseases of plants are in fact abiotic and caused by adverse weather conditions. For example, a grower reported a large patch of potatoes were healthy in the morning, but a few hours later had turned completely black and collapsed. After eliminating other factors, it turned out there was a severe thunderstorm earlier that morning–the plants were hit by lightning and electrocuted!

Information on these basic environmental factors may not always be available. For example, if you have just purchased a plant, flowers, or fruit that is diseased there may not be any available environmental data for you to examine, so you will need to look at the host directly.

4 Host

It is worth checking the host plant for signs of physical damage that could act as an entry point to opportunistic pathogens. Wounds can be caused naturally through strong winds, or animals grazing, or culturally, for example pruning or strimmer damage. Identify and note the approximate age of the plant. Young plants are often more prone to diseases than more established ones. It is now time to look at the symptoms themselves. If possible try and discover when the symptoms first appeared: was it pre- or post-planting? Do you have any evidence that the seed was disease free? Using Figure 1.7, examine each part of the plant (check it all, including the roots) and record the type of symptom observed and its location. In most cases, many different symptoms may be expressed by the plant, depending on the stage of the pathogen's life cycle and the plant physiology. For example, an early symptom of infection of *Zymoseptoria tritici* in wheat is water soaking of patches of leaves. This is followed by chlorosis and necrosis. Later in the season the necrotic patches often possess small 'dots' which are its fungal fruiting bodies. Certain diseases, for example *Xylella fastidiosa*, and many viral diseases of plants require a vector, so look for high infestations of insects in the locality of the diseased plant. We have to be careful when taking samples–it is always prudent to take a number of samples to test in the laboratory. Choose samples showing a range of symptoms from what appear to be early mild symptoms through to severe (remember the warning earlier in this chapter). If possible you need to provide a non-infected plant for comparison (e.g. some plants have dark vascular tissue that could be mistaken for a disease symptom if a healthy plant was not available for comparison).

Examination and identification

Before you can begin to treat a plant disease, or take steps to prevent its spread, you must identify the pathogen involved. This is a complex process, which has benefited greatly from the development of new molecular approaches such as gene sequencing. There are a number of different techniques which can be used, both in the field and in the laboratory. These methods will be considered in more detail in Chapters 2–4, with each chapter focusing on specific types of pathogens and different techniques.

Scientific approach 1.1

Koch's postulates

Classically, establishing **pathogenicity** is achieved by completing a series of procedures that establish four criteria, referred to as Koch's postulates. These were formulated in the 1880s by the medical bacteriologist, Robert Koch while investigating anthrax, but have subsequently become accepted for other pathogens. The steps are:

1. The suspected causal agent must be consistently associated with the disease symptoms.
2. It must be isolated from the diseased plant and grown in pure culture.
3. When a pure culture of the suspected causal agent is inoculated into a healthy, susceptible plant, the host must reproduce the same disease symptoms as those that were originally observed.
4. The same causal agent must be isolated from the freshly diseased tissue and must have the same characteristics as the organism that was first isolated.

Establishing Koch's postulates is an important way of determining pathogenicity in humans—see *Human Infectious Disease and Public Health* by William Fullick (OUP 2019) in this series of primers.

The same process can be applied in plant diseases. Although for many plant pathogens establishing Koch's postulates is possible, some obligate pathogens require a host plant to grow and therefore cannot be isolated onto a growth medium such as an agar plate. In these cases, step 2 of Koch's postulates may need to be skipped and direct inoculation of a healthy host may be necessary. Completion of Koch's postulates is always required before a new host/pathogen can be officially described.

Pause for thought

Following an outbreak of a new disease, how would you prioritize understanding the epidemic?

Why is establishing Koch's postulates important?

Positive impacts of plant disease

Although diseased plants generally have a negative impact on the economy, environment, or social well-being, a number of plant pathogens are in fact encouraged, providing delicacies for different cultures. One example is the case of corn smut (*Ustilago maydis*), where the fungal spores infect maize ovaries at flowering resulting in swelling (hypertrophy) of the corn kernels, which eventually turn black as the spores mature. These are particularly regarded as a delicacy known as Huitlacoche (sleeping excrement!) in Mexico, where they are picked and used in quesadillas or soups (Figure 1.13). Grapes may be purposely infected with a form of *Botrytis cinerea* that results in shrivelled, mouldy grapes. However, the sweetness of the grapes during this process is intensified considerably as they dehydrate and they take on new flavours due to the production of a compound known as phenylacetaldehyde which is also found in cocoa. The wine produced (named Noble Rot) is sold at a premium price throughout the world. And during the seventeenth century a craze for exotic tulips caused financial boom—and bust (more of this in Chapter 4). Plant diseases, like the plants they affect, are a constant thread running through our social history.

Figure 1.13 Huitlacoche (sleeping excrement) from corn smut is a delicacy in Mexico

Bernardo Ramonfaur/Shutterstock.com

Chapter summary

- Plant diseases are all around us and affect our economy, environment, and social well-being.
- They are the result of a pathogenic interaction with a susceptible plant by a fungus, protist, bacterium, or virus.

- Plants respond in different ways to an infection resulting in expression of symptoms and signs in all or some of its leaves, stems, roots, flowers, and fruit.
- Classification, taxonomy, and nomenclature are human constructs identifying similarities and differences between causal agents of plant disease, grouping and naming accordingly.
- In order for a plant disease to survive in any particular environment there is a requirement for a susceptible host, suitable organism, and the pathogen itself.
- Epidemiology investigates the where, why, when, what, and how of plant disease.
- Understanding the ways in which plant diseases are spread is key to their management, control, and prevention.
- Diagnosis of a plant disease problem requires a holistic approach and requires gathering knowledge of the influence of host, environment, pathogen, and interactions by humans.

Further reading

Agrios, G.N. (2005) *Plant Pathology*. Academic Press. 952 pp.

The most comprehensive book on plant pathology published to date.

Buczacki, S. and Harris, K. (2014) *Pests, Diseases and Disorders of Garden Plants*. 4th Edition. William Collins. 512 pp.

A good basic overview of common plant pest and disease.

Burchett, S. and Burchett, S. (2017) *Plant Pathology*. Garland Science. 244pp.

A recent general overview of plant pathology.

Greenwood, P. and Halstead, A. (2018) *RHS Pests & Diseases: New Edition, Plant-by-Plant Advice, Keep Your Produce and Plants Healthy*. DK. 224 pp.

A good overview of common diseases found in your garden, along with good management practices.

http://www.apsnet.org/Pages/default.aspx

The American Phytopathological Society plays an important role in coordinating and sharing information on plant diseases and pathogens, and publishes the journal Phytopathology.

https://www.bspp.org.uk/

British Society of Plant Pathology. This charitable organization covers all aspects of UK plant pathology, runs training courses and conferences, and publishes two scientific journals.

Discussion questions

1.1 How could the impact of a new plant disease be reduced?

1.2 What questions would you ask a grower who had a problem with their plants?

1.3 How would you decide the type of microorganism that has caused a plant disease?

1.4 How would you prioritize studying aspects of a plant disease epidemic?

2 FUNGI AND FUNGUS-LIKE ORGANISMS

Dr Paul Beales

The fungi and protista kingdoms contain a wonderful, diverse group of organisms comprising of somewhere between 1.5 and 5 million species, although only around 100 000 species have been described. Of these, only about 10% are actually plant pathogens—the remainder live as saprophytes or in a symbiotic relationship with their host plant (e.g. mycorrhizal fungi).

This chapter will explore some of the best-known plant pathogens in these two commonly accepted kingdoms. An overview of some taxa within these kingdoms is shown in Table 2.1.

Table 2.1 Overview of protista and fungi kingdoms (based on Kirk *et al* 2001)

Kingdom	Phylum	Common examples
Protista	Oomycota	*Phytophthora*, *Pythium*, downy mildews
Fungi	Chytridiomycota	*Synchytrium*, *Olpidium*
	Zygomycota	*Rhizopus*, *Mucor*
	Ascomycota	Powdery mildews, *Glomerella*, *Venturia*
	Basidiomycota	Rusts, smuts, fly agaric mushroom

Some key differences between the kingdoms fungi and protista

Fungi are eukaryotic, heterotrophic organisms that depend upon external organic carbon sources broken down by secreted fungal enzymes and absorbed as dissolved molecules by the fungus. In other words, they cannot make their own food by photosynthesis as plants do, so they secrete enzymes which break down and digest other organisms, either alive or dead, outside of the body of the fungus. The products of digestion are then absorbed by the fungal cells. You will see later that some specialized fungi obtain their food by associating themselves so closely with the host cells that they almost become part of it. This way, the plant is kept alive, but the parasitic fungus obtains its food from nutrients that should have been the host plant's!

Fungi have chitin in their cell walls (compared to cellulose in the cell walls of plants or oomycetes). They reproduce by spores and their vegetative growth state is generally in the form of hyphae. They can be single-celled, like yeasts, or multicellular, like the fly agaric toadstool in Figure 2.1.

Figure 2.1 This fly agaric mushroom (*Amanita muscaria*) is the sort of thing that comes to mind for most of us when we think of fungi—but they are only the tip of the iceberg

Anthony Short

At first sight, some protist plant pathogens are quite similar to fungi—which is probably why they were classified as fungi for years! Unlike fungi, they don't have chitin in their cell walls. Another key distinguishing feature of these fungus-like organisms is that they have a motile phase in their life cycle, and move in water using one or more flagella which act like an outboard motor, propelling them to new food sources (Figure 2.2). The broader group includes some algae, and some members

Figure 2.2 The spore cells of fungi and Protista show clear differences under the microscope which cannot be seen by the naked eye

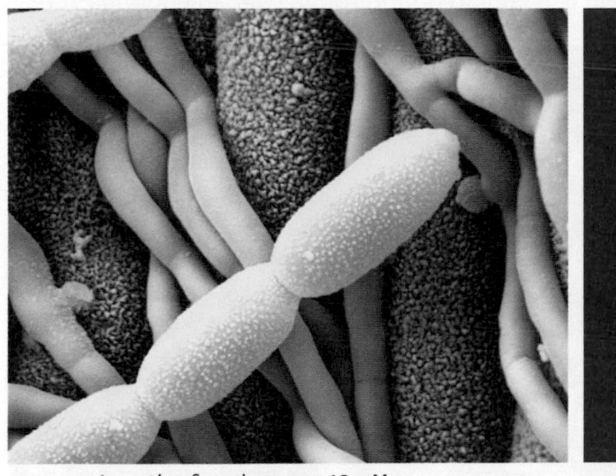

—— Length of each spore 10 μM

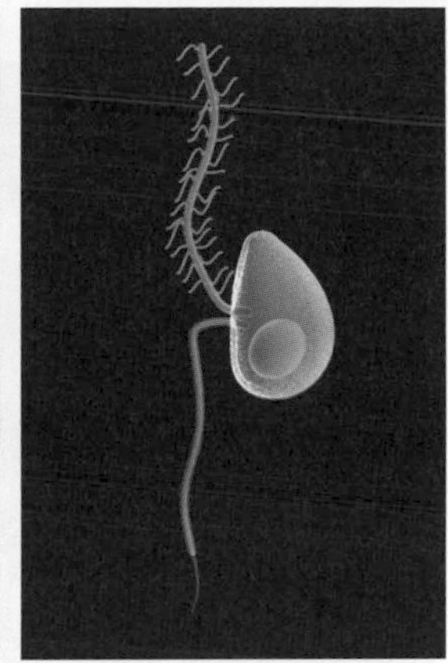

—— Height of spore (top to bottom) 10 μM

Biophoto Associates/Science Photo Library

are photosynthetic, although the protista that act as plant pathogens are all heterotrophic.

Both fungi and protista are spread locally, via rain splash or wind carrying their spores, over longer distances via animals or contaminated soil, for example attached to vehicles and shoes, or over enormous distances of thousands of miles via plant trade.

The members of these groups that act as plant pathogens are usually micro-fungi, and often require specialist tools such as microscopes to see the necessary features for identification, as you can see in Figure 2.2. One exception is *Armillaria mellea*, commonly known as honey fungus. This is one of the most prolific plant pathogens in the UK. It has a wide and varied range of host plants including viburnum, eucalyptus, and apple, all common garden plants. Honey fungus fruiting bodies are more typical of mushrooms that are frequently observed in the autumn (see Figure 2.3), but the main body of the fungus is the huge mycelium which spreads through the soil in which they grow, interconnecting plants. In fact, an *Armillaria* fungus in Oregon, USA, is thought to be the largest living organism on Earth, and possibly the oldest as well. It spreads across about 2.4 miles, with a surface area of 3.7 square miles, and it is around 8000 years old!

In the next section of this chapter you will discover some of the most commonly encountered and impactful plant diseases caused by fungi and protista. You will see an indication of the defining features of the groups, any specialist characteristics, examples of their hosts, the impact on both plants and people—and what to look for to recognize, identify, and diagnose an infected plant.

Figure 2.3 The fruiting bodies of honey fungus may look harmless—but the majority of the fungus is microscopic and hidden underground or associated with its host plant. It attacks and kills a wide range of woody and perennial plants and can spread underground over an enormous area.

Anthony Short

Oomycetes—fungus-like members of the protista

The main taxonomic group within the kingdom protista is called the oomycetes. Plant diseases caused by this group have changed the course of human history—and still have the potential to do the same today. Three primary groups of parasites spring to mind when discussing the oomycetes, namely the phytophthoras, pythiums, and downy mildews.

Phytophthora

Ask a group of people whether they could name a plant pathogen, and if they can think of one at all, the vast majority will mention potato blight 'fungus'. Unbeknown to most people, the pathogen which causes potato blight is not a fungus at all, but an oomycete (*P. infestans*). Phytophthora derives from Greek for 'plant destroyer', and many *Phytophthora* species live up to this grand name (see Table 2.2).

The number of described Phytophthora species discovered has increased over the years (see Figure 2.4). The number of species being identified is increasing at an exponential rate as more research is performed and more sensitive detection and diagnostic methods are discovered. Currently over 100 species have been described.

Phytophthora symptoms

Symptoms of a Phytophthora infection can appear on any part of a plant as a water-soaked lesion or rot (see Figure 2.5). Infections generally occur

Table 2.2 Some of the most damaging species of Phytophthora

Main Phytophthora species			Impact
P. megakarya, P. palmivora	Cocoa		Average loss 10–20% of world's annual production. In wet years, losses can be up to 80%.
P. palmivora	Coconut		Killed in excess of 11 000 coconut palm trees in the Philippines in 1989–1990. This is a vital crop that supports 3.4 million farm families in the Philippines alone.
P. capsici	Black Pepper		Pepper is a major export commodity for Indonesia. In the Lampung area the complete crop was lost in the 1930s, and in 1965 there was a 32% crop loss in other areas.
P. meadii, P. botryosa, P. palmivora	Rubber		In parts of Asia, 10% of cultivated area may be affected; in severe cases 40% yield loss was calculated.
P. infestans	Potato		1.5 million people killed due to starvation in Ireland in the 1850s. Untreated potato yield loss following a recent 25-year study was 10.1 tonnes/hectare.
P. nicotianae	Citrus		Present in numerous citrus groves in Florida and Brazil causing substantial root rot to 8–20% of the citrus plants.
P. cinnamomi	Eucalyptus		Large forest areas wiped out by this pathogen in parts of Australia.
P. ramorum	Larch		Sudden oak death (USA), millions of trees killed in Northwestern states. In the UK, large areas of larch woodlands affected in SW England and Wales—this forest crop no longer planted in these regions.

Cocoa: Anthony Short; Coconut: Anthony Short; Black Pepper: Anthony Short; Rubber: tanewpix/Shutterstock.com; Potato: Anthony Short; Citrus: Anthony Short; Eucalyptus: janaph/Shutterstock.com; Larch: Julia Skoryk/Shutterstock.com

Figure 2.4 Graph to show the rate of discovery of species of Phytophthora affecting different plants since 1860 (courtesy of Clive Brasier, Forest Research)

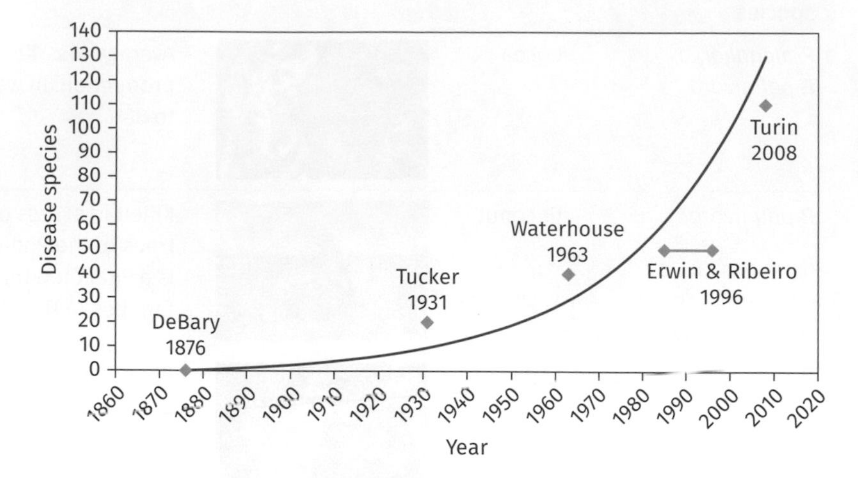

Figure 2.5 Phytophthora can destroy a potato plant in a very short time if conditions are right. Example shows potato blight.

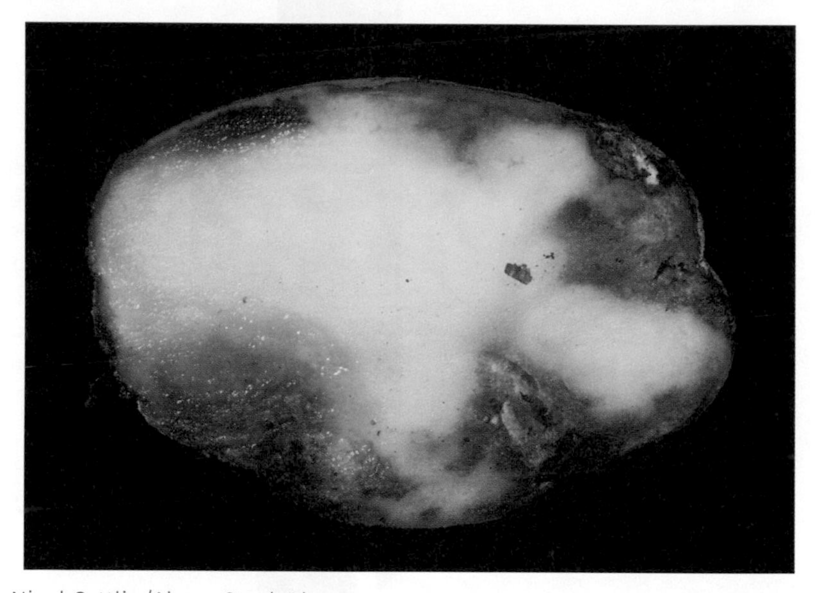

Nigel Cattlin/Alamy Stock Photo

in one of two regions of a susceptible plant (with some exceptions), either underground, i.e. in the roots, tubers, and lower stem, or above ground, i.e. leaves, shoots, flowers, or fruit. This is primarily due to subtle differences in the asexual, sporing structure, the **sporangium**, and its method of spread. Root and stem base decay can lead to wilting (resembling drought), and in severe cases death. Leaf infections do not usually kill the plant, but may look awful, lead to dieback, and impact growth and fruit/seed development.

Host range

Phytophthora affects almost all plant families (with the exception of many grasses) to varying degrees. A recent study carried out in the UK by the Royal Horticultural Society, based on samples sent into their diagnostic clinic, found 25 different species of phytophthora affecting 87 types of plants from 50 different families.

Key life cycle and epidemiological features

The hyphae of phytophthora have no cross walls, which makes them different to the fungi. They also have a variable number of sporing stages including resting spores, asexual spores, and sexual spores.
The main reason phytophthoras are such successful pathogens is that they have a range of ways in which they can survive difficult conditions and spread from plant to plant.

- **Chlamydospores:** These asexual survival spores form when conditions become unfavourable for growth. The outer wall becomes thickened and often darkens (see Figure 2.6). The spore is then protected against UV radiation, drying out, and attack by other microorganisms. In this form phytophthora can survive decades in the soil until a susceptible host and favourable conditions return. Many chlamydospores form during the winter months, as deciduous trees drop their leaves.
- **Zoospores:** The most prolific spores of phytophthora are zoospores. They form in special bodies called sporangia and they have flagella, so they can move about. Around 500 sporangia can form per cm^2 of plant tissue and 10–30 zoospores form in each sporangium—no wonder

Figure 2.6 This image shows phytophthora in survival mode—note the darker colour of the spores and the thick cell walls compared to the lighter hyphae

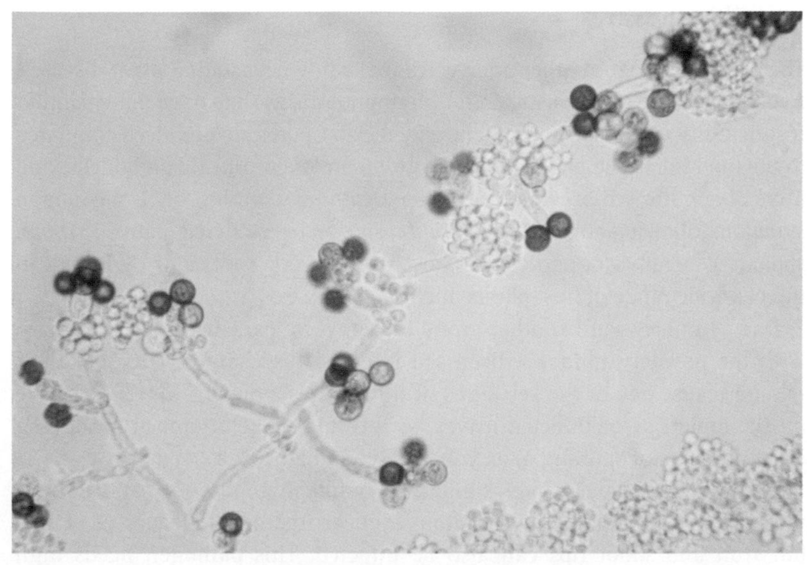

Scale: Diameter 50 μM

Gado Images/Alamy Stock Photo

Figure 2.7 This image shows zoospores that have encysted on a leaf, germinated, and are growing (note the germ tubes) towards a stoma

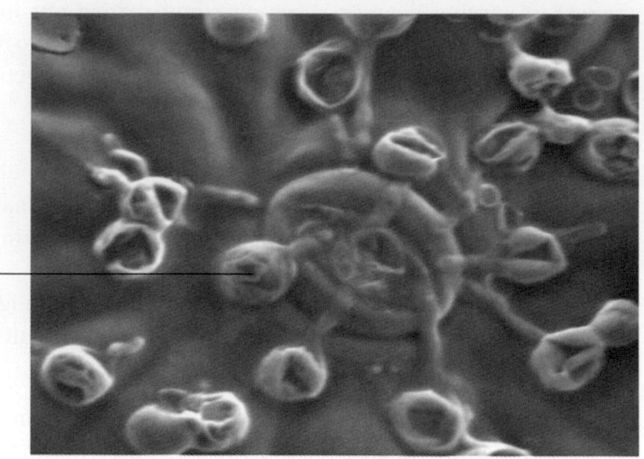

phytophthoras are very infectious! The zoospores need water to spread and don't live long. They swim in water courses, or in droplets on leaves or roots and they are attracted to natural openings (e.g. stomata), where they encyst (lose their flagella) and germinate (Figure 2.7).

- **Oospores**: These are produced by sexual reproduction, when special 'male' and 'female' structures unite. The oospores which are produced have very thick walls and can survive for many years in soil and dead plant material. They may be more virulent or tolerant to fungicides due to recombination of DNA from separate mating types.

Pythium

There are a number of other oomycetes that cause devastating plant diseases. *Pythium* spp. are often mistaken for phytophthora, as they have many similar features, differing only subtly in the way the zoospores are produced compared to phytophthora. The majority of this group are opportunistic pathogens, and often attack stressed or wounded plants. Death by **damping off** is a common symptom following an attack by *Pythium* spp. on overwatered plants. *Pythium* spp. are generally not particularly host-specific–one species of *Pythium* can infect a wide range of host plants, including grasses.

Downy mildews are another group of oomycete parasites, often confused with the powdery mildews discussed later in this chapter. They get their name because one of the key signs of infection is a downy-like appearance on the underside of infected leaves, as a result of the organism producing huge numbers of sporangia, as you can see in Figure 2.8.

The parasite shares many similarities with phytophthoras and control methods often overlap. The symptoms appear first on the leaves of plants, but fruit and shoot tips can also be infected. This pathogen needs high humidity to complete its life cycle and so downy mildews are more common on closely packed plants growing undercover. This means they can be

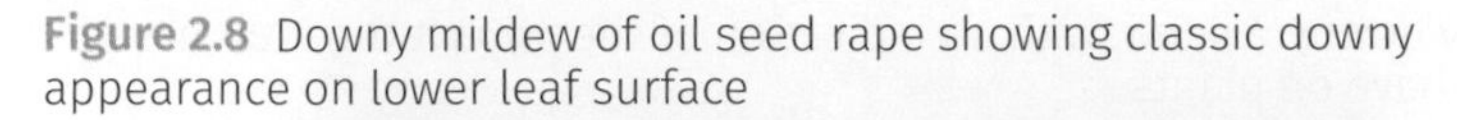

Figure 2.8 Downy mildew of oil seed rape showing classic downy appearance on lower leaf surface

Nigel Cattlin/Alamy Stock Photo

devastating to crops grown in greenhouses and polytunnels. Downy mildews affect a range of key vegetable, horticultural, and fruit crops, some of which you can find in Table 2.3.

Fungi

The fungi are a group of organisms of enormous complexity, ranging from single-celled yeasts to large, multicellular organisms like the honey fungus (see Figure 2.3). They play a vital role in the ecosystems of the world as decomposers, and as mycorrhiza, which are symbiotic relationships between fungi and plants. However, fungi can also wreak havoc, devastating growing crop plants, and also destroying crops once they have been harvested and stored. For many years scientists were unsure even how to classify fungi—look at Scientific approach 2.1 to find out more.

How do fungal plant diseases spread?

The primary method of spread of fungi is by spores. They make a huge range of different types of spores as a result of asexual or sexual reproduction. Some rely on passive methods of spread, and others rely on an active approach where the parent organism has built-in mechanisms to discharge spores—look out for some of them as you read this chapter. Passive methods of spore dispersal include wind, water droplets, contact with infected material, and remaining dormant in the soil until a suitable host is planted nearby. As a result, fungi can be difficult organisms to control.

Table 2.3 Examples of crops affected by downy mildew, with an indication of the impact they can have on yields, showing the impact they can have on plants

Downy Mildew	Primary Host	Impact
Pseudoperonospora cubensis	Cucumber 	30–40% yield losss on untreated infected crop.
Plasmopara obducens	Impatiens 	Seed-borne disease that almost wiped out the common Busy Lizzie (*Impatiens × walleriana*), in the UK when controls failed in 2011.
Plasmopara viticola	Grape 	Yield loss 50–80% under favourable conditions as this disease attacks young flowers and fruit.
Plasmopara halstedii	Sunflower 	Serious disease of many countries, affecting oil yield up to 95%.
Peronospora manshurica	Soybeans 	Yield loss of up to 12% in untreat-ed crops.

Cucumber: spicyPXL/Shutterstock.com; Impatiens: Anthony Short; Grape: © Getty; Sunflower: kurhan/Shutterstock.com; Soybeans: Vacheslav Rubel/Shutterstock.com

Scientific approach 2.1
Fungal names

Nomenclature, or the process of naming organisms, has a long a and varied history. Fungi have proved to be one of the most complex and often confusing kingdoms to name, as traditionally there can be more than one name for any given fungus and they can change frequently following advances in **morphological** and molecular taxonomy.

The International Code of Nomenclature for algae, fungi, and plants (ICN) ensures a uniformity in naming everything from plants to fungi. There are some basic rules that are followed whenever a new species is named:

- The given name must be a Latin binomial (following the code's protocol).
- The name must be published.
- It must be validated with a concise description, including illustrations where possible of the diagnostic features of the new taxon. The description should include date of collection, locality, and the substrate or host.
- Following the naming process, the type specimen should be deposited in a national herbarium.

How many names can a fungus have?

All officially identified plants have one binomial name, although they may have a common name as well; for example a common oak tree is officially *Quercus robur.* However, in the unique case of fungi, different states of the fungus (based on asexual or sexual spores found) can have different names. There is currently a move to ensure all newly discovered fungi have just one Latin binomial, but you will still see older descriptions with two very different names!

Example 1:

When a disease commonly referred to as anthracnose was first described, only the asexual (haploid) stage—the **anamorphic stage**—was observed, and the fungus was given the name *Colletotrichum gloeosporioides* and classified in the old 'bucket group' Fungi imperfecti or Deuteromycota. This fungus affects a broad range of tropical, subtropical, and temperate plants such as cereals, fruits, and tree leaves, causing leaf and fruit spots and blotches, as you can see in Figure A.

Sometime later, the sexual (diploid) sporing stage of the fungus—the **teleomorphic** stage—was discovered, and the fungus was reclassified into the group Ascomycota *and* given a new or 'true' name *Glomerella cingulata.* General good practice states that once the teleomorphic name is described it takes precedence over the anamorph, even if only the asexual form is seen on a diseased plant. However, both names are still valid, so scientists will often use both. This inevitably causes a lot of confusion, even more so when

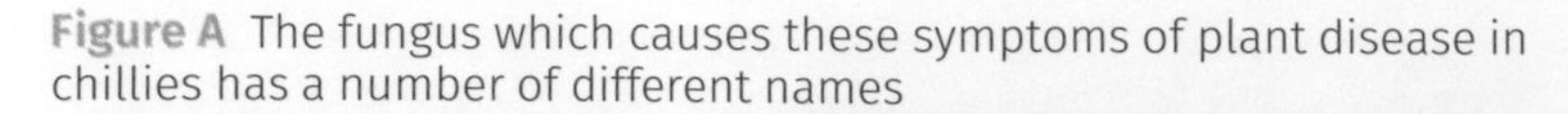

Figure A The fungus which causes these symptoms of plant disease in chillies has a number of different names

DAwee/Shutterstock.com

taxonomists reclassify fungi and give them a further two new names! Only the most recent names are actually valid. These new names are based on advances in knowledge of the organism, for example by molecular diagnostics.

To address these issues, in 2012 the 'one fungus one name' system was proposed at the ICN congress. This proposal supports the abolition of dual nomenclature. In most cases the sexual name takes precedence; however, dependent on committee agreement, sometimes the asexual state which is widely known may take precedence, and the above is one such example, where *Colletotrichum* will be the accepted name over *Glomerella* (Canon 2012). To bring some order to the complications this system would invoke, for names published prior to 1 January 2013 a choice had to be made and only one of them recorded. To make things even more complicated, this can be either the anamorph or teleomorph name, depending on a rule of 'first named takes precedence'. After this date, only a single name for a new fungus is valid. The rules do sometimes take some interpretation and best practice usually involves a committee discussion around a particular name rather than rigid adherence to rules!

Within taxonomy, the name described above is generally the lowest ranking given, and states the grouping into a genus and species (although there are some exceptions, e.g. sub-species or special forms). As more and more features are included in the system, the groups become larger and are given other generic names, for example family all the way up to kingdom level. An example of how cereal powdery mildew is taxonomically ranked is shown in Table A.

The preferred scientific name of a fungus includes a post nominal authority, or the name/abbreviation and date of the individual who first published the organism's name, for example using the example above, *Blumeria graminis* (DC) Speer (1975). If the fungus has been reclassified by a different taxonomist over the years, all are included, so the first recorded descrip-

Table A Classification of cereal powdery mildew

Kingdom	Fungi
Phylum	Ascomycota
Subphylum	Pezizomycotina
Class	Leotiomycetes
Order	Erysiphales
Family	Erysiphaceae
Genus	*Blumeria*
Species	*Blumeria graminis*

Powdery mildew of cereals is ranked according to special unique features for each group.

tion of this fungus was by Augustin Pyramus de Candolle (1778–1841), a well-known botanist of his time (his name is abbreviated to DC) who named the mildew *Erysiphe graminis* back in 1815. The fungus was then reclassified by Speer in 1975 to its current name, *Blumeria graminis*.

A species name is usually only written in full once in a text—after the first writing the species is abbreviated, for instance the example given would be written *B. graminis*.

With advances in ultrastructural, biochemical, and molecular biological studies, fungal names are regularly under review. No doubt there will be more changes in the future!

Pause for thought

Imagine you have discovered what appears to be a new fungal plant pathogen. In a paragraph describe what you might need to consider before giving it a name.

Types of pathogenic fungi

As you will have realized by now, the fungi are an enormous, complex kingdom of organisms. The pathogenic fungi are found in a number of different groups—a brief summary of each group, along with examples of diseases caused by members of the group, follows.

Chytridiomycota

Commonly referred to as chytrids, they have chitinous cell walls, and are the only group of fungi with motile spores, but unlike protist plant

Figure 2.9 Cauliflower-like swellings on potato tuber caused by the chytrid *S. endobioticum*—and one of the hardy spores which make it so hard to eradicate the disease

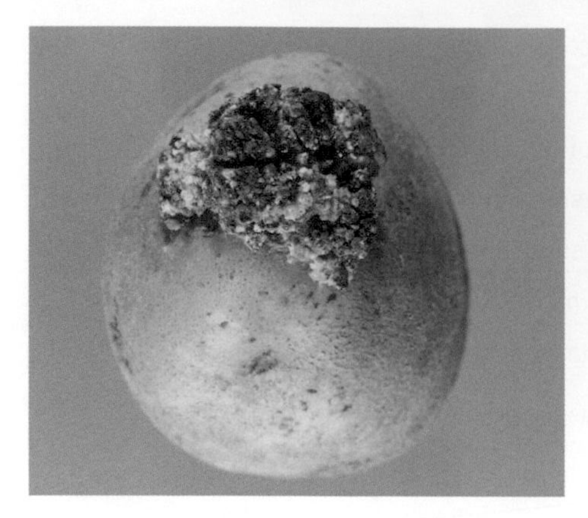

Crown Copyright

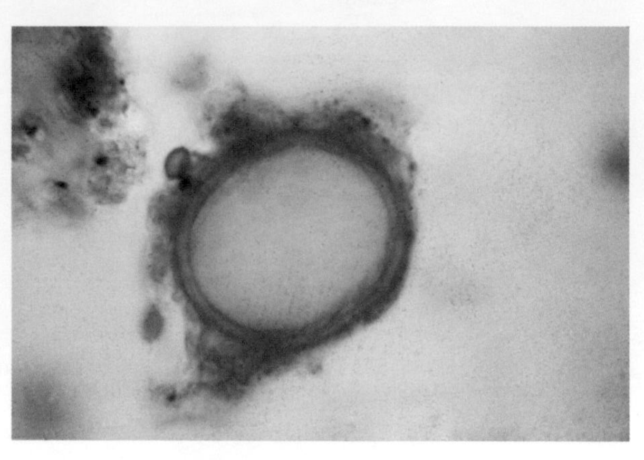

UK Crown Copyright - courtesy of Fera

pathogens they only possess a single flagellum. They are classified as fungi based on various ultrastructural features and their cell wall chemistry. There are very few plant pathogens in this group; however, a notable example is wart disease of potato caused by the chytrid fungus *Synchytrium endobioticum*. This disease was thought to have been introduced to Europe in the aftermath of the 1850s potato blight disaster. It spread to nearly all potato-growing countries in Europe, until statutory measures and development of 'immune' varieties finally restricted its spread. Disease symptoms occur mainly only on the potato tuber (the bit you eat!). The symptoms appear as warty, cauliflower-like swellings as you can see in Figure 2.9. As this fungus has a motile sporing stage, the disease requires wet soils for local spread, and over longer distances by movement of infected seed potatoes or via contaminated soil attached to farm machinery. An interesting fact about this fungus is its ability to survive in soils—the resting spore stage (in Figure 2.9) can survive over 30 years and still affect a crop! If this disease is found, a grower faces severe restrictions on growing further potatoes, until the land can be tested and proven free from potato wart.

Zygomycota

These include fungi you are likely to find growing on your strawberries, or bread that has been left out a little too long. The vast majority of these fungi are saprophytic (the fungus feeds on dead plant material) and opportunistic in nature. They are so common that they are often referred to as mycological weeds! The group is classified due to the formation of resistant spores (zygospores) following the sexual joining of two hyphal strands. However, most people are familiar with the asexual spores that form; as in Figure 2.10, they resemble black-headed 'pins' in the affected substrate. Each 'pinhead' (sporangium) can contain many thousand spores, and the support (sporangiophore) is positively phototropic and negatively

Figure 2.10 Pin mould on tomato

Alex Hyde/Science Photo Library

gravitropic. This means they grow strongly upwards, ensuring that the spores are dispersed as far as possible by wind, water drops, or physical touch. Figure 2.10 also shows you an example of a zygomycete affecting a yellow courgette. The pathogen enters openings caused by insects or other physical damage and can result in up to 75% crop loss.

Ascomycota

The Ascomycota are by far the largest group of the fungi, containing over 64000 species. The group contains the fungal part of lichens, saprophytes, and plant parasites and includes both macro and micro-fungi. The word 'asco' is Greek for sac or wineskin. Sexual spores (ascospores) are formed within an ascus and this is the key distinguishing feature for this group. Table 2.4 gives some notable examples from the many plant pathogens within the ascomycota.

Powdery mildews

Unlike the downy mildews described above, powdery mildew fungi are a unique group within the Ascomycota, classified in the taxon order *Erysiphales*. They affect a broad range of crop and horticultural plants from grasses to trees—over 10000 species in all. Most have a relatively narrow host range, with some exclusive to a particular species. For example, the wheat strain of *Blumeria graminis* will not infect barley, and the barley strain of the same powdery mildew will not infect wheat.

The name 'powdery mildew' is given to these fungi as they produce copious amounts of asexual spores, that *en masse* give a powdery appearance to affected plants, as you can see in in Figure 2.11. Powdery mildews are obligate pathogens, keeping their host alive whilst feeding. Asexual spores germinate on the host substrate (usually on the leaves), grow a little, then

Table 2.4 Plant pathogen examples from the Ascomycota

Plant pathogen	Common name	Host example/s	Impact
Blumeria graminis	Powdery mildew	*Triticum aestivum, Hordeum vulgare* (wheat, barley)	Worldwide, one of the worst foliar diseases of cereals. Uncontrolled yield losses vary, but can be up to 60%, although more commonly 4–9%.
Ophiostoma novo-ulmi	Dutch elm disease	*Ulmus* spp. (elm)	20 million trees (70% of the UK population) killed in a decade following introduction of the fungus.
Cryphonectria parasitica	Chestnut blight	*Castanea* spp. (sweet chestnut)	3.5 billion trees killed in USA since accidental introduction from Asia. First found in UK in 2011.
Venturia inaequalis	Apple scab	*Malus* spp. (apple)	Affects yield, fruit quality, and storage longevity. Up to 14 fungicide applications may be needed per season to manage disease.
Magnaporthe grisea	Rice blast	*Oryza sativa* (rice)	Global food insecurity caused by this fungus. 30% of global rice production losses—enough to feed 60 million people!
Taphrina deformans	Leaf curl	*Prunus* spp. (peach, almond)	Attacks on *Prunus* fruit crops in the USA accounts for 2.5–3.0 million dollars' loss.
Glomerella cingulata	Anthracnose	*Mangifera indica* (mango)	One study in Nigeria showed 60% of all mango trees were infected and 34% fruits were unmarketable.
Claviceps purpurea	Ergot	*Secale cereale, Triticum aestivum* (rye, wheat)	Ergotism, holy fire, or St Anthony's fire resulted in tens of thousands of people losing limbs, toes, and fingers, or dying in the eleventh century CE after eating rye bread contaminated with ergots.
Diplocarpon rosae	Rose blackspot	*Rosa* spp. (rose)	Impacts quality and marketability of plants. Hemibiotroph—Total leaf loss may result following a severe infection. The fungus often overcomes host resistance and up to 40 sprays per season are needed to manage the disease.

(Continued)

Table 2.4 Plant pathogen examples from the Ascomycota (*Cont.*)

Plant pathogen	Common name	Host example/s	Impact
Mycosphaerella musicola	Sigatoka of banana	*Musa* spp. (banana)	Often found as a complex with other pathogens, regarded as the most serious problem for bananas. Yield losses up to 50%, and premature ripening prevents exports.
Gaeumannomyces graminis	Take-all	*Triticum aestivum* (wheat)	Estimated that half of UK wheat crops are affected. Average yield loss 5–20%. Cannot be controlled with pesticide. Crop rotation is key to control.
Monilinia fructicola	Brown rot	*Prunus* spp. (e.g. cherry, peach)	Losses of peaches due to infection can be up to 75%.
Hymenoscyphus fraxineus	Chalara ash dieback	*Fraxinus* spp. (ash)	Significant economic impact on ash-growing industry across Europe (further information in case study).
Gibberella zeae (anamorph name *Fusarium graminearum*)	Fusarium head blight	*Triticum aestivum*, *Hordeum vulgare* (wheat and barley)	Several species of *Fusarium* cause head blight. Fungus produces a toxin (mycotoxin) that is highly carcinogenic to animals and humans.
Zymoseptoria tritici	Septoria blotch	*Triticum aestivum* (wheat)	One of the most important foliar diseases of wheat. Can cause up to 20% loss in untreated crops.

Case study 2.1

Ash dieback in Europe—the importance of a name

Ash dieback was first recorded back in the early 1990s in Poland. The disease slowly spread across other countries in Europe, but at the time it was unclear what was causing the infection. The disease will kill young saplings within a year or two, but more mature ash trees can take up to 15 years to die. Each year, unaffected side shoots emerge from the main stem, which in time themselves become infected and die back. Often more mature trees die from a combination of the weakening effect of Chalara ash dieback and other pathogenic organisms such as honey fungus (Figure 2.3). The sexual stage of this fungus was believed to be the ascomycete *Hymenoscyphus albidus*, a common saprophyte fungus found across continental Europe and Great Britain. When a pathogen is indigenous and established in a country, it is impossible in practical terms to implement statutory action to prevent the spread of the disease. Statutory action involves putting a legal framework in place designed to contain and eradicate the disease and gather evidence to inform future policy. However, further molecular analysis on the fungus in 2011 revealed it to be a brand new species, named *H. pseudo-albidus* because it was similar to, but distinct from, the saprophytic strain. It was proved to be a pathogen using Koch's postulates—see Chapter 1. Later, the name was further changed to *H. fraxineus* (T. Kowalski) (Baral, Queloz & Hosoya 2014).

As this newly identified fungus had not been previously recorded in the UK when it was first identified, statutory action could be taken. In order to establish the extent of infection, an ambitious, unprecedented survey was organized, bringing together numerous organizations to check ash trees in every 10 km^2 of the UK for the presence of this disease. The survey took considerable organization of field staff, and testing in two UK laboratories was scaled up to manage the influx of thousands of samples. Results came in thick and fast, and within one week the whole of the UK had been surveyed, so knowledge of the extent of the pathogen could be given to ministers, who were then able to make informed disease management decisions. The survey has continued to the present date, and disease progression can be seen in Figure A.

The life cycle of *H. fraxineus*

The fungus produces wind-borne, sexual spores (ascospores) which land on ash leaves and infect them. The fungus grows into the leaves, and down into the woody shoots, killing tissue in its wake. The pathogen causes significant purplish-coloured cankers on the trunk, and cross sections show a dark brown stain typical of infection, as you can see in Figure B. Dead, infected leaves remain longer than uninfected leaves on the tree, but eventually fall to the ground, where the fungus overwinters in the rotting leaves. The following year, when the new leaves have formed on the tree, the fungus emerges from the fallen leaves as sexual spores in small fruiting bodies, or apothecia, which are carried by the wind to infect fresh leaves.

What hope is there for the future? At the moment, no one really knows. But some scientists are trying to identify the genetic factors which make some ash trees tolerant of the fungus, or able to resist the disease. If this happens, we may be able to grow more resistant trees and plant healthy ash trees for future generations to enjoy.

? Pause for thought

- The number of pathogens infecting UK trees seems to be increasing alarmingly. Some people blame global warming. Look into these claims and list any evidence you discover for or against this theory.
- Tracking the spread of a disease like ash dieback, and then investigating ways of producing resistant trees, costs a lot of money. Discuss the value of this type of work—is it money well spent? Justify your opinions.

Figure A East-to-west development of *H. fraxineus* following regular surveys

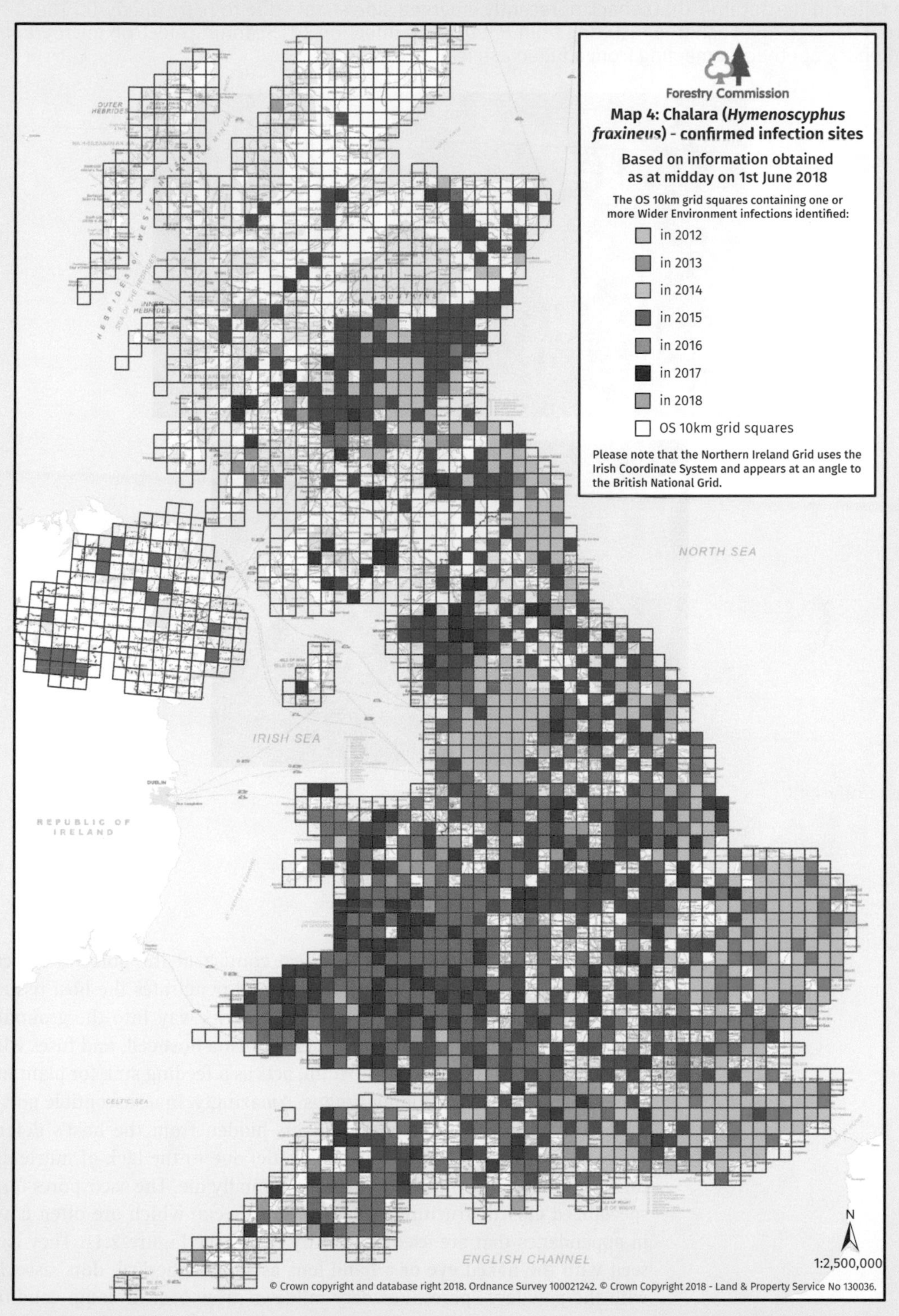

Forestry Commission

Figure B (a) Ash leaves affected by *H. fraxineus*—they remain on the trees long after the healthy leaves have fallen in the autumn. (b) Dieback of recently emerged side shoots due to *H. fraxineus*. (c) The internal damage to an ash tree resulting from *H. fraxineus* infection. (d) Scanning electron micrograph of ash dieback apothecia emerging from a fallen ash leaf (×20).

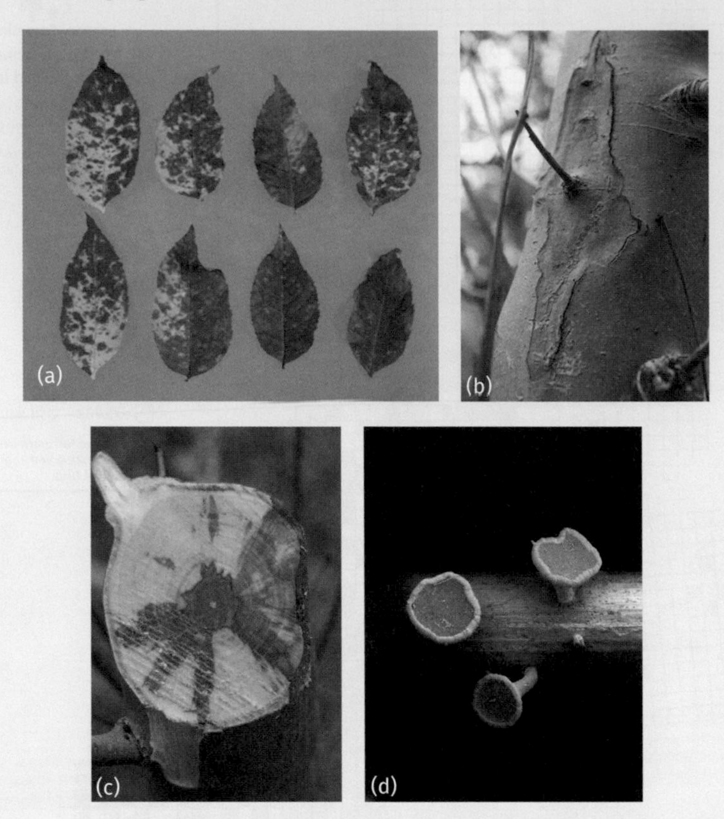

produce a firm infection 'base' in direct contact at the point of infection. An infection peg emerges from this base and penetrates the host tissue directly (a little bit like an oil rig drill working its way into the ground). As the fungus continues to grow, it presses against a host cell, and fuses with it. This structure, known as a haustorium, acts as a feeding sink for plant nutrients, to nourish the developing fungus. Amazingly, in a susceptible powdery mildew infection, the pathogen remains hidden from the host's defences. Infected leaves often turn yellow (chlorotic) due to the lack of nutrients (as the fungus has taken them!), and may eventually die. The ascospores form in specialized closed structures called chasmothecia, which are often covered in appendages that are used to identify them—see Figure 2.11. They can be seen with the naked eye or a hand lens as small spherical 'dots' associated with the powdery spores. Powdery mildews grow best in damp conditions, but they can tolerate much drier conditions than most fungi.

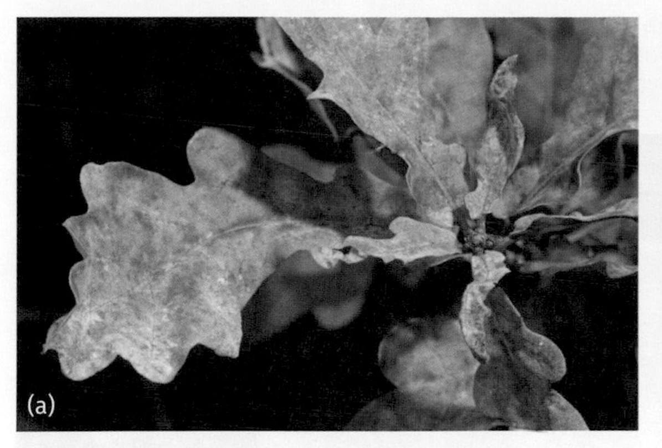

(a) Holger Kirk/Shutterstock.com;

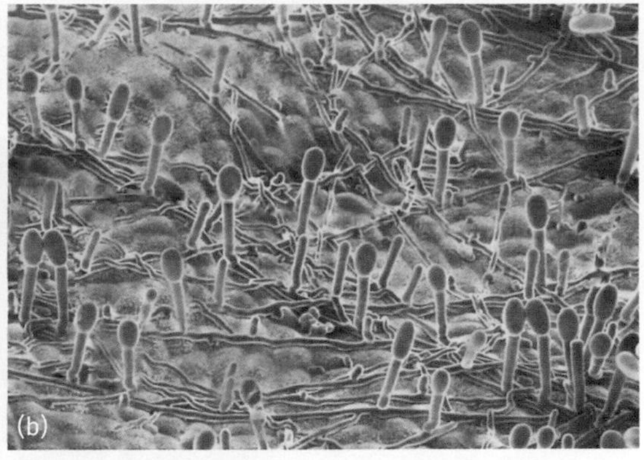

(b) Crown Copyright

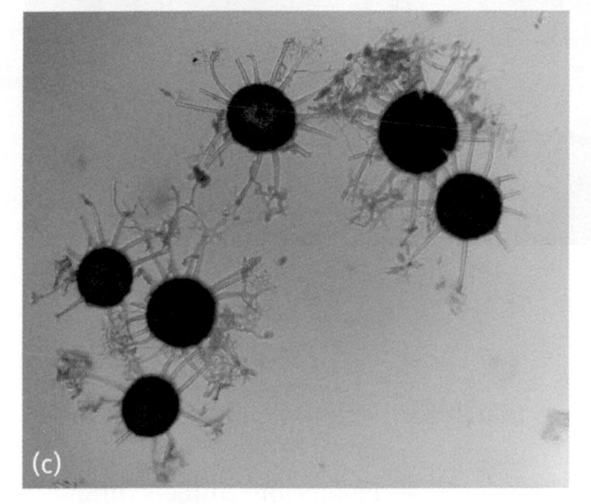

(c) Crown Copyright

Figure 2.11 Here is an example of powdery mildew on oak leaves, showing the powdery fungus on the leaves and close-ups of the asexual (b), and sexual © fruiting bodies. These are used together to help scientists diagnose powdery mildews.

Basidiomycota

The majority of mushrooms (macro-fungi) that you might see in the autumn—or buy in the supermarket—fall into this group. With a few exceptions, these fungi are primarily saprophytes or symbionts (e.g. *Amanita muscaria*, fly agaric—see Figure 2.1—has a mutualistic relationship with *Betula* spp. trees, including the common silver birch). The size of the fruiting body is, however, not the defining feature for this taxon. Specialized club-shaped cells called basidia that support basidiospores (external sexual spores) on short extensions called basidiophores are what scientists look for to identify the Basidiomycota.

Rusts, smuts, and bunts are the most well-known plant pathogens found in the Basidiomycota and we will look briefly at each of these in turn.

Rusts

Of all the fungi, the rusts have the most impressive life cycles. Some rusts can produce up to five different spore stages and share more than one host plant, and the signs of the pathogen can be extremely variable. All rusts are obligate plant pathogens, and therefore require a living host to complete

Figure 2.12 This bramble leaf, photographed in the autumn, displays two different types of spores—the bright rusty summer spores and the black over-wintering survival spores

Anthony Short

their life cycle. Unlike some fungi that infect stressed plants, rusts infect healthy, vigorously growing plants.

For example, bramble rust (*Phragmidium violaceum*) infects bramble plants. Brambles are incredibly common, very prickly plants which produce brambles (also known as blackberries). The fungus completes all five stages of its life cycle on the same plant. If you turn over a bramble leaf during the summer, you may observe a bright to burnt orange (rusty!) coloration emerging from the lower leaf surface (Figure 2.12). Closer examination reveals these are spores of the fungus erupting through the leaf surface (known as rust pustules). If, however, you turn the same leaf over in the winter, the spores have changed and are now black. These are in fact a completely different sporing stage (and not just the old orange spores observed earlier). The black rust pustules contain overwintering 'survival' spores.

Black stem rust (*Puccinia graminis*) is a long-term enemy of the human race. It particularly affects cereals and over centuries has caused global widespread crop loss. The romans even named a god, Robigus, after the fungus. They sacrificed a dog annually in an effort to save their crops from rust—but it didn't really help anyone (particularly the dog!). There are not many cereal crops at the right stage of growth in the winter, so in order for this obligate pathogen to survive, it moves to an alternative non-annual host for the winter months. This alternative host is a plant called barberry. In the USA, a campaign was launched to eradicate barberry, so the rust fungus that devastated the cereal crops would not be able to complete its life cycle. However, due to the fact that barberry grows almost everywhere in the USA, this was only partially successful.

During the 1950s the impact of black stem rust was reduced because scientists produced rust-resistant wheat plants by selective breeding. However, during the late 1990s, severe epidemics of this rust were once again seen in Uganda, and then began to spread to other areas. The fungus had overcome the resistance gene and this new strain, named Ug99 (initially isolated in Uganda (Ug) in 1999 (99)), resulted in a resurgence of the pathogen. This fungus is still a problem in various parts of the world and work is continuing towards its management (see Chapter 7 for more on this story).

Smuts and bunts

Smut (of Germanic orgin, meaning 'dirt') and bunt fungi are pathogens of monocotyledonous plants—primarily grasses, but also others such as sugarcane and maize. They produce copious amounts of dark, thick-walled, dusty spores that form in place of the normal grain/fruiting structure. Examples include corn smut (Figure 2.13; described in Chapter 1 as a delicacy in parts of the world), *Ustilago hordei* of barley, and *U. nuda* of oats.

Figure 2.13 Severe corn smut, where each kernel develops into a mass of black smut spores

Natalia Melnychuk/Shutterstock.com

Bunt fungi, classified in the genus *Tilletia*, are smuts where the spores are formed in a modified papery membrane known as a bunt ball. Bunt fungi include a serious contaminator of grain, *T. caries* (common bunt or stinking smut). The dark spores completely replace individual wheat grains. Infected crops may be partially stunted, but the primary impact is that bunted grain spores have a rotten, fishy smell. When the grain is harvested

Figure 2.14 Did you know that the origin of gingerbread men was directly due to stinking smut infection? A European baker was unable to sell bread as the wheat had been infected with stinking smut. He therefore masked the flavour (and the greyish colour) by mixing it with ginger and molasses and the gingerbread man was born!

Kenneth Dedeu/Shutterstock.com

the bunt spores 'flavour' the remaining grain and anything made from it (see Figure 2.14).

Fungi imperfecti or Deuteromycota

When a fungus is discovered only in its asexual state it is unknown where it should fit into the classification scheme and is therefore put into the artificial (officially unrecognized) category of *Fungi imperfecti* or Deuteromycota (also sometimes referred to as mitosporic fungi). It is a sort of holding area until further analysis can classify the fungus more correctly. Some plant pathogen examples that are still held in this group are *Alternaria* spp. (e.g. *A. solani*, which causes early blight of potato) and *Verticillium* (e.g. *V. albo-atrum*, a wilt-causing disease of a range of crop plants).

Detection methods for fungi and fungus-like organisms

We cannot diagnose many plant diseases, or control or eliminate fungal and fungus-like pathogens, if we cannot identify them. Detection of fungal and fungus-like pathogens may involve direct visual examination of symptoms, looking for signs of infection, and methods to aid detection, including inducing fungal sporulation. On some occasions there is a clear relationship between symptoms and signs, such as leaf spotting and the presence of rust pustules or white powdery spores of a powdery mildew on the upper leaf surface. Frequently, however, further testing is required to develop cryptic (hidden) infections or techniques to extract the primary parasite from water, complex media such as soils, or direct from a plant that is rotten and contaminated with saprophytic, non-pathogenic bacteria and fungi. The use of molecular and serological techniques discussed throughout this book has, over recent years, become more commonly used, and detection of fungi regularly involves use of such methods. You will learn more about these in later chapters. But traditional detection methods still have their place, and should be considered as they are simple, fit-for purpose, and cost-effective and permit screening of a wide range of pathogens. These are the techniques we will focus on for fungal diseases in this chapter.

Incubation and isolation

Fungi and fungus-like organisms favour high humidity to produce their sporing structures, and this is the primary form by which such organisms are identified. Following a preliminary examination of the plant with a hand lens or dissecting microscope and recording of symptoms, you may not be able to see fruiting bodies—for example, if it is a young infection. If you place the plant material in a damp chamber (e.g. a clear plastic box containing damp kitchen roll) and leave it for a while (usually from a few days to a couple of weeks), this will encourage the fungus to produce its spores. Microscopic examination of the incubated material using dissecting microscopes (which magnify up to about 50×) and compound microscopes (up to 1000× magnification) allows you to record fine morphological details. You can record these and compare them against defining characters for the suspected taxon. If you can't complete identification following incubation, isolation onto a semi-selective medium may be required. You can do this directly from the incubated material, although the risk of contamination by opportunistic and saprophytic fungi and bacteria is significant. It is better to isolate the pathogen directly from the symptomatic plant. It is essential to choose appropriate material for your culture. The leading edge, or the outer limit where symptoms can be seen on the plant, gives the best chance of isolating the primary pathogen (see Figure 2.15). If you sample too far either side of the leading edge you may miss the pathogen completely, or grow a saprophytic opportunistic organism that is feeding on the tissue killed by the primary pathogen.

Figure 2.15 The leading edge of the fungal infection on this leaf is easy to identify

Preto Perola/Shutterstock.com

Top tip: Using a small square of sticky tape can remove key fungal parts *in situ* from leaf tissue and is particularly useful for examining visible signs of infection, such as powdery mildew or rust fungi. Press a $1cm^2$ piece of tape gently onto the affected area, remove, and place onto a microscope slide containing a drop of water or stain. The tape also serves as a cover slip, and the slide can be viewed directly under a compound microscope.

Isolation method

First use a surface disinfectant to remove any contaminants. Then use sterilized tools such as a scalpel and forceps to cut small pieces of tissue from the leading edge and place them onto a growing medium in a petri dish. The medium (a semi-selective medium) can have variable amounts of nutrients and antibiotics that favour or 'select' certain groups of pathogens over others. Some media contain almost no nutrients and these encourage sporulation of the more stubborn pathogens. Incubate your culture for a set period of time at a specific temperature, and then examine your plate both macro- and microscopically for diagnostic features of the fungus. You can see the results of some of these cultures in Figure 2.16.

Baiting for fungal pathogens

Baiting is a particularly useful tool to detect fungal pathogens from what can be difficult substrates, such as soil and water, where you cannot see the fungus. Bait plants consist of known susceptible hosts for fungi or fungus-like organisms you are checking for. For example, as mentioned earlier, *Phytophthora* spp. require a water phase as part of their life cycle and rivers/water courses can be good areas to test for such pathogens. A common bait method uses a 'BOB' or 'bag of baits' and consists of small pieces of mixed plant material wrapped in muslin cloth and tied off (see Figure 2.17). This is placed in the test water for three days, and *Phytophthora* zoospores are attracted to the baits, enter through the pores in the muslin cloth, and

Figure 2.16 Fungal cultures

Scott Bauer/US Department of Agriculture/Science Photo Library

Figure 2.17 A 'bag of baits' is a simple but effective way of testing for fungal and fungus-like pathogens in water

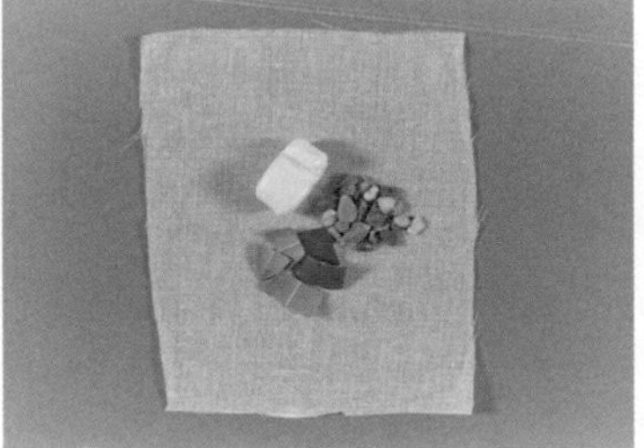

Crown Copyright

infect the plant pieces. The BOB is then removed and you can either use a specific device to test directly for *Phytophthora*, or the material can be surface sterilized and grown in culture.

Spore trapping

Many fungal pathogens, such as rusts and mildews, produce airborne spores, and a wide range of air-sampling devices have been developed to capture them. They generally fall into passive or active categories. Passive traps can be as simple as a sticky rod that spores adhere to as they pass. Active traps draw air carrying the spores into the device using, for example, a suction motor (see Figure 2.18). Spores captured on the traps can then be identified by visual and molecular methods, or used in epidemiological studies.

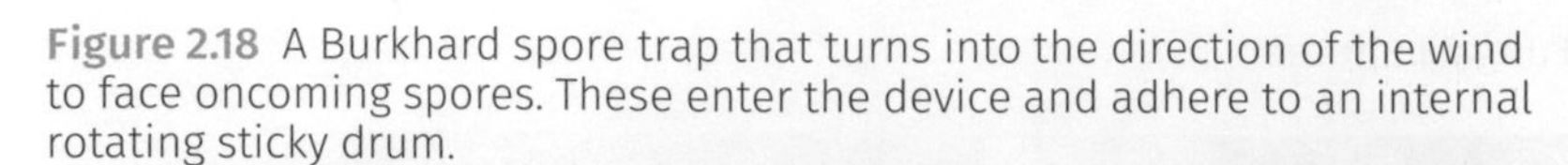
Figure 2.18 A Burkhard spore trap that turns into the direction of the wind to face oncoming spores. These enter the device and adhere to an internal rotating sticky drum.

Chapter summary

- Fungi are eukaryotic, heterotrophic organisms which have chitin in their cell walls and reproduce by spores. Protista are quite similar but they have a motile phase in their life cycle, with spores that move in water using flagella.
- *Phytophthora, Pythium*, and downy mildew plant diseases are very efficient at spreading, infecting a wide range of plants, and surviving adverse conditions. They all require damp/wet conditions to complete their life cycle.
- Fungi may have more than one name, depending on what spore stage has been observed or new features discovered. Changes in the naming rules are gradually coming into effect which will simplify this system into one fungus, one name.
- Fungi are a diverse, versatile, and adaptable group of organisms that populate the entire world. Plant pathogenic fungi are generally microscopic; however, their devastating effects on plants can be clearly seen. Some feed whilst keeping their host alive; others kill their host before feeding. The wide range of sporing structures ensure these organisms are efficient at spreading and surviving in adverse conditions.

- We have a wide variety of modern techniques available to use for identifying fungal pathogens, but some of the classical culturing and microscopic methods still remain as effective approaches. Many microorganisms will be found on any single plant affected by a disease that needs to be identified. One of the skills of a good plant pathology diagnostician is to eliminate these from the primary causal agent.

Further reading

https://www.bspp.org.uk/

British Society of Plant Pathology. Lots of great resources that can be downloaded.

https://www.britmycolsoc.org.uk/

British Mycological Society. As above, lots of great resources covering all aspects of fungi. Often organize fungal forays where you can find mushrooms and toadstools in woodlands.

http://www.apsnet.org/Pages/default.aspx

American Phytopathological Society. Numerous resources available. Publish the journal Plant Disease.

Lane, C.R., Beales, P. and Hughes, K.H. (2012) *Fungal Plant Pathogens (Principles and Protocols).* Principles and Protocols Series, Cabi Publishing Cabi International, Wallingford, Oxfordshire, UK. 304pp.

Discussion questions

2.1 Discuss the range of jobs/work areas that require consideration of fungal plant diseases.

2.2 Imagine you were a plant pathogenic fungus. Give yourself a name and prepare a successful battle plan to attack and infect a plant group of your choice (on a local/national or worldwide scale). Think about how and where the plants are grown and how you might successfully spread, how you and your fellow fungi will get into the plant and survive, and what you would do if you were successful in your fight.

2.3 A distressed grower calls you and wants to discuss an issue they are having with their prized plants. Think about the key questions you would need to ask to help diagnose what is wrong. Discuss what advice you would give to the grower to manage or control the problem. Revisit this question after reading the rest of this book.

3 PLANT PATHOGENS: BACTERIA

Dr John Elphinstone

Bacterial plant diseases affect a wide range of crops, ornamentals and environmentally important plants, including trees. Since the first descriptions of bacterial plant diseases by the early pioneers in plant bacteriology, many different genera of bacteria which damage plants have been reported (see Figure 3.1 for examples of bacterial diseases recognized for many years). Bacterial plant diseases have an enormous global impact, in both human and financial terms. Economic losses are difficult to quantify, but they are usually quoted in terms of tens to hundreds of millions of US dollars annually around the world. Types of losses can range from minor blemishes, such as spots, areas of dead tissue (**necroses**), or yellowing on leaves or fruit, to more serious rots, wilts, blights, and diebacks, which can spread through and devastate entire crops or areas of natural vegetation. Even minor symptoms can be economically significant if the loss in quality affects marketability, for example for high-value ornamental plants or food crops. People in developed countries often will not buy fruit or vegetables which look anything less than perfect.

Losses from bacterial diseases may be reflected in lower yields, because diseased plants simply do not produce as much as healthy crops. However, more serious impacts are usually felt when bacterial diseases are found in specimens which are part of the national and international trade of plants. The problem is at its worst when the plants are involved in breeding programmes and people want to use them for propagation, or to produce seeds. When this happens, whole consignments can be rejected and may lose all value, which can be devastating to the growers.

If a bacterium is discovered for the first time in an area or country, quarantine regulations may require that the whole consignment is destroyed and that expensive disinfection measures are put in place to prevent further spread. Since compensation is rarely paid in such cases, the cost of this kind

Figure 3.1 Fireblight of pear was first described by T.J. Burrill in 1880 and bacterial wilt of tomato by E.F. Smith in 1896. These bacterial diseases still attack our plants today.

of loss is usually met by the trader, or even the grower who was originally contracted to produce the consignment. They may further suffer from a damaged reputation and loss of future business.

There are virtually no effective or approved chemical bactericides available to control bacterial plant pathogens, and there are very few examples of plant varieties with effective resistance against them. As a result, the control of bacterial plant diseases is largely dependent on being able to successfully diagnose bacterial infections at an early stage and taking precautions to prevent their introduction and spread in the first place, as you will see in Chapter 7. In this chapter we will take a closer look at the **phytopathogenic bacteria**—bacteria which cause diseases in plants. We are going to look at the types of disease they cause, the methods used for their classification and identification, and some of the most important pathways via which they can spread and cause disease in their **host plants**.

Phytopathogenic bacteria

The great majority of bacteria are beneficial to animals or plants, or have no impact on their lives; only a small proportion are pathogenic. All of the plant pathogenic bacteria fall into only three of the 30 currently recognized bacterial phyla: (1) Gram-negative Proteobacteria; (2) Gram-positive Actinobacteria; and (3) cell-wall-lacking Tenericutes. Gram-negative and Gram-positive bacteria are identified by the way a compound called teichoic acid in their outer layers takes up a coloured dye known as Gram stain (Figure 3.2). This was developed in 1884 by a Danish pathologist named Hans Christian Gram, who accidentally struck on a way of easily dividing many bacteria into two distinct groups. This technique is still used today.

Most Gram-positive and Gram-negative phytopathogenic bacteria are single-celled, rod-shaped bacilli of 1–2 μm in length, which do not form endospores. Endospores are a survival mechanism used by some bacteria

Figure 3.2 Micrographs of Gram-positive (right) and Gram-negative (left) stained phytopathogenic bacteria

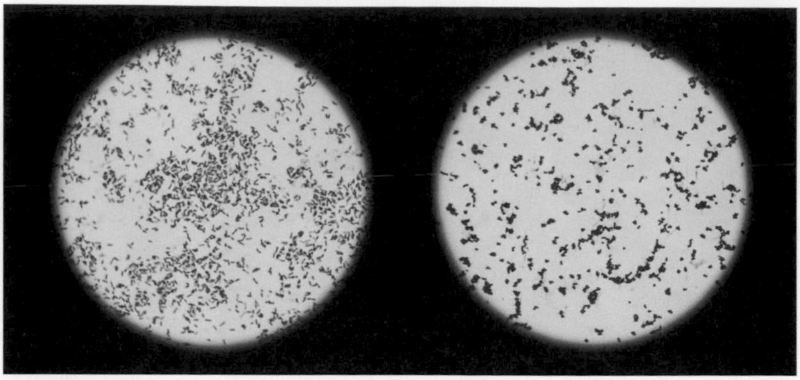

toeytoey/Shutterstock.com

when conditions are harsh. They form tough walls around them which can survive for years. As you can imagine, if phytopathogenic bacteria could form endospores, they would be particularly difficult to deal with. Some phytopathogenic bacteria have one or more flagella and some don't. Exceptionally, some of the Actinomycetales (such as *Streptomyces* species) are filamentous bacteria that produce well-developed vegetative strands called hyphae. These can grow to 0.5–2.0 μm in diameter with aerial branches that develop exospores at the end. They look rather like incredibly tiny moulds!

We all learn that bacteria have cell walls—but in fact, there are some specialized groups that manage without them! The bacteria which make up the spiroplasmas and phytoplasmas of the Tenericutes phylum are characterized by a lack of cell wall; amazingly, they are essentially free-living protoplasts with tough cell membranes containing sterols or lipoglycans.

These specialized bacteria live in protected, osmotic-neutral habitats in the gut or haemolymph of their insect vectors, or in the phloem of their host plants. We can culture spiroplasmas in bacteriological growth media in the lab, whereas we cannot persuade phytoplasmas to grow. The spiroplasmas

Figure 3.3 The scanning electron microscope reveals some of the amazing corkscrew shapes of phytopathogenic spiroplasma bacteria

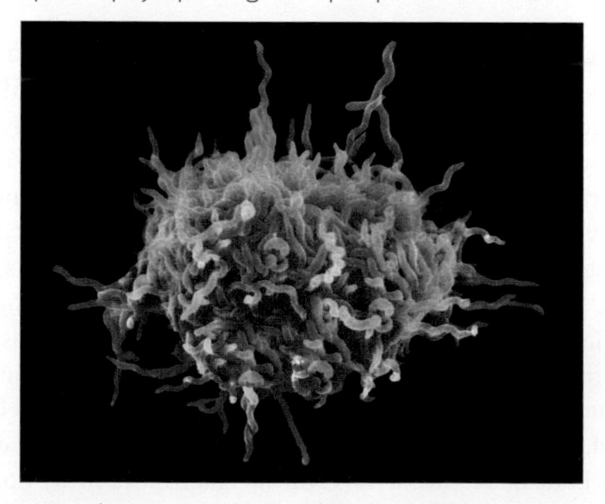

Dr David M. Phillips/Visuals Unlimited/Getty Images

have spiral-shaped cells and move in a corkscrew, motion whereas phytoplasmas have lots of different shapes ranging from spherical to long and thin, or even filamentous. (As such, we say their *cellular morphology* is variable.) Figure 3.3 shows you an example of a spiroplasma.

Case study 3.1

Fire blight disease—an example of why bacterial plant diseases matter

A walk through the fruit and vegetable a isles of any supermarket demonstrates the popularity of apples and pears. They are eaten fresh and cooked in so many ways, from a crunchy apple munched fresh from the tree, to apple pies, poached pears—and, of course, cider and perry! Yet the orchards that produce these wonderful fruits are at risk from a deadly bacterial infection known as fire blight disease.

Fire blight disease is caused by the Gram-negative bacterium *Erwinia amylovora*. It mainly affects apples (*Malus* spp.) and pears (*Pyrus* spp.), but it has a range of other hosts, including hawthorn (*Crataegus* spp.), firethorn (*Pyracantha* spp.), quince (*Cydonia* spp.), cotoneaster (*Cotoneaster* spp.), and mountain ash (*Sorbus* spp.), many of which grow wild in our hedgerows. This makes it especially difficult to deal with the disease because it is found in wild populations of plants which cannot be controlled.

E. amylovora is an example of an insect-vectored bacterial plant pathogen—it is spread by pollinating insects such as honey bees, and

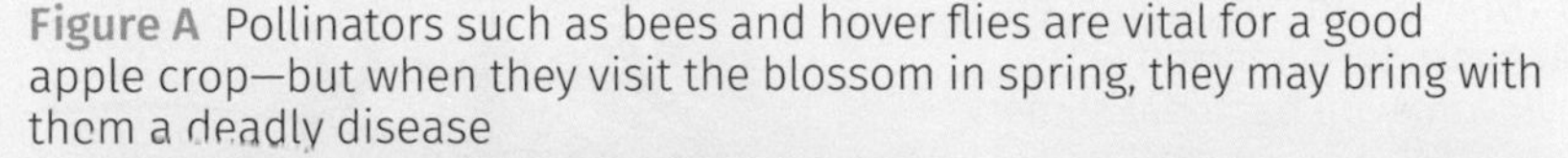

Figure A Pollinators such as bees and hover flies are vital for a good apple crop—but when they visit the blossom in spring, they may bring with them a deadly disease

Anthony Short

initially infects the host plant in the spring through the flowers (Figure A). The bacteria multiply in the high-sugar environments on the stigma and in the nectaries, penetrating through the natural openings in the floral cup. We can notice the first symptoms of fire blight disease when infected flowers shrivel and die, becoming brown or black. Infection initially spreads as more insects visit the flowers; the spread can continue later in the season by rain splash to the leaves, where the bacteria can enter through open stomata or wounds created by heavy rain, hail, or insect feeding.

The infections eventually spread extensively through the host tissues of the whole tree, causing die back of shoots, which turn from brown to black from the tip. They characteristically bend near the tip to form a so-called 'shepherds-crook' shape. Infected leaves and fruits appear blackened and tend to remain attached. Eventually, whole branches (or even whole trees) completely die back.

Necrosis of the older infected branches leads to the formation of cankers at the limits of infection. Cankers are open wounds where the cork cambium and wood has been destroyed, surrounded by callus tissue which forms as part of the plant's defences. The bacteria survive the winter season in the margins of such cankers and multiply in the following spring, leading to exudation of bacterial slime, which can then be further spread by rain splash and visiting insects or even birds. You can see some of these symptoms in Figure B.

Fire blight disease is native to North America and was introduced to Europe in the 1950s, since when it has spread through most European countries. One interesting feature of its spread was its movement along railways and roads, where hawthorn hedges were planted. Losses have been extensive, particularly in pear orchards. For example, in the main pear-growing area in Italy, this disease resulted in the destruction of over one million trees be-

Figure B (a) Photo of twig dieback with 'shepherds-crook' formation at tip. (b) Infected pear florescence (c) Hawthorn twig infected with *E. amylovora* with bacterial slime exudate.

Figure C Destruction of infected trees by burning.

UK Crown Copyright—courtesy of Fera

tween 1994 and 2004, either directly due to the disease or due to deliberate removal and destruction of orchards to prevent its further spread. Several susceptible pear varieties—for example, the delicious Laxton's Superb—have been lost completely.

The most favourable environmental conditions for fire blight infection are temperatures from 18 to 29°C, high relative humidity (90–95%), and wet plant surfaces, for example following rain. During the flowering period of the tree, temperatures as low as 12°C can still result in infection. Knowledge of the exact conditions that favour disease development has allowed the development of fire blight forecasting systems, which are computer-based systems that help orchard growers to predict disease risk by monitoring the weather conditions (e.g. temperature, rainfall, and leaf wetness).

There is no cure for fire blight infection. When an orchard is infected, the only option is to destroy the tree, as in Figure C. The best hope is to control the spread of the disease. The main control measures to prevent introduction of fire blight into an orchard are to source new planting material from fire-blight-free areas. The movement of beehives from areas where fire blight occurs to disease-free areas is also regulated.

Pause for thought

Although fire blight has already spread extensively throughout Europe, there are still areas which remain free from the disease in most countries. How important do you think these areas are in terms of restricting further losses in our pear and apple orchards? What steps would you take to ensure that they remain free from *Erwinia amylovora* in the future?

Classifying and identifying the pathogens—old ways and new

In a similar way to the pathogenic fungi you discovered in Chapter 2, the Latin names of plant pathogenic bacteria seem to have been constantly changing as the methodology used for their classification, nomenclature, and identification has developed, allowing more accurate determination of their taxonomy and phylogenetic relatedness. Today it is accepted that many species, belonging to some 30 different genera of bacteria, are known to cause disease in plants and trees (Table 3.1).

Originally, bacteria were classified according to the morphology (appearance) of their cells when observed under the microscope, or according to the colour, shape, consistency, and even the smell of the colonies they form on different microbiological growth media. Over the years more detailed classification systems have developed. These involve comparing the physiological and biochemical properties of the different bacterial pathogens at the cellular level, in a process known as polyphasic taxonomy. This involves the use of multiple approaches to observe differences between bacterial phenotypes, often including combinations of the following:

- growth temperatures (maximum, minimum and optimum);
- ability to grow in different salt concentrations;
- susceptibility to a range of antibiotics;
- ability to utilize different sources of carbon (e.g. sugars, alcohols, or organic acids) and nitrogen (amino acids) to grow or to produce acid or alkali by-products;
- activity of bacterial enzymes: using tests to detect the formation of end-products typically produced by different bacteria or by observing the activity of enzymes that directly hydrolyse their substrates (such as proteins, polysaccharides, or fats (lipids));
- reaction in different serological tests to antibodies which specifically recognize differences in cell wall components (such as glycoproteins) or enzymes and polysaccharides secreted by different bacteria;
- susceptibility to a range of viruses known as bacteriophages, each able to infect and kill different bacteria;
- profiles of the types and amounts of cellular fatty acids, unique to different species of bacteria, measured in the laboratory using gas chromatography mass spectrometry (GCMS; see Scientific Approach 3.1) of the fatty acid methyl esters (FAMEs);
- profiles of cellular proteins and peptides, separated according to their differential movement in an electric field through a polyacrylamide gel in a process called polyacrylamide gel electrophoresis (PAGE).

As you can imagine, the classification or identification of a plant pathogenic bacterium can be a long process involving a lot of testing and observations that could take months to complete.

Table 3.1 Plant pathogenic bacteria, with the main symptoms they cause

Phylum	Class	Order	Family	Genus	Main symptoms
Gram-negative bacteria					
Proteobacteria	Alphaproteobacteria	Rhizobiales	Rhizobiaceae	'*Candidatus* Liberibacter'	Stunting, greening
				Rhizobium	Galls, hairy roots
		Sphingomonadales	Sphingomonadacea	*Sphingomonas*	Corky roots
	Betaproteobacteria	Burkholderiales	Comamonadaceae	*Acidovorax*	Blight
			Burkholderiaceae	*Burkholderia*	Rots
			Burkholderiaceae	*Ralstonia*	Vascular wilt
			Oxalobacteraceae	*Herbaspirillum*	Leaf stripes
	Gammaproteobacteria	Enterobacteriales	Enterobacteriaceae	*Enterobacter*	Cankers, dieback
				Erwinia	Rots
				Pectobacterium	Rots
				Brenneria	Rots
				Dickeya	Rots
				Pantoea	Vascular wilt
				Samsonia	Cankers, dieback
				Serratia	Vascular wilt
		Pseudomonadales	Pseudomonadaceae	*Pseudomonas*	Leaf spots, galls, cankers, wilts
				Rhizobacter	Galls
		Xanthomonadales	Xanthomonadaceae	*Xanthomonas*	Leaf spots, cankers, vascular wilt, blight
				Xylella	Leaf scorch
				Xylophilus	Necrosis and cankers

Table 3.1 Plant pathogenic bacteria, with the main symptoms they cause (*Continued*)

Phylum	Class	Order	Family	Genus	Main symptoms
Gram-positive bacteria					
Actinobacteria	Actinobacteria	Actinomycetales	Micrococcaceae	*Arthrobacter*	Vascular wilt
			Microbacteriaceae	*Clavibacter*	Vascular wilt, stunting
				Curtobacterium	Vascular wilt
				Leifsonia	Stunting
				Rathayibacter	Gumming (producing lots of sticky gum), stunting, distortion
			Nocardiaceae	*Nocardia*	Galls
				Rhodococcus	Galls
			Streptomycetaceae	*Streptomyces*	Scab
Tenericutes	Mollicutes	Acholeplasmatales	Acholeplasmataceae	'*Candidatus* Phytoplasma'	Virescence, phyllody, flower sterility, shoot proliferation, internode elongation, yellowing
		Entomoplasmatales	Spiroplasmataceae	*Spiroplasma*	Stunting, internode shortening, distortion, yellowing, mottling

Scientific approach 3.1

Fatty acid profiling

Fatty acids are an important part of the structure of the cell membrane of microorganisms. The types and quantities of fatty acids are unique to different species of bacteria and can therefore be used to identify them. The fatty acid profile of an isolated unknown bacterium can be determined using a technique called gas chromatography mass spectrometry (GCMS). The main stages of this process are summarized below:

1. The bacteria are isolated and cultured under standard conditions (e.g. by streaking on **trypticase soy agar** and incubating for 24 hours at 28°C).

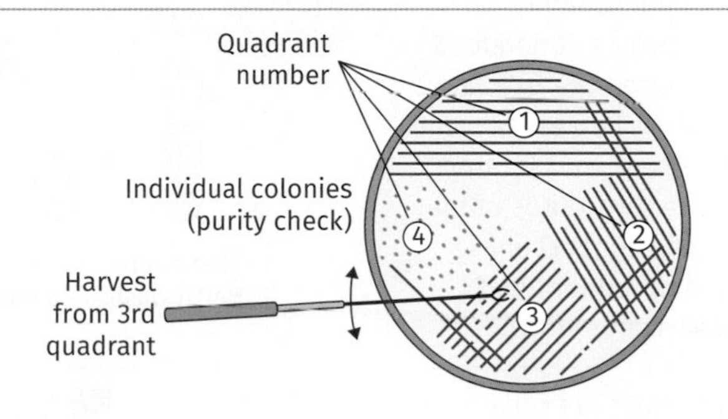

2. Bacterial cells are harvested by removing ~40 mg of culture from growth on agar plate.

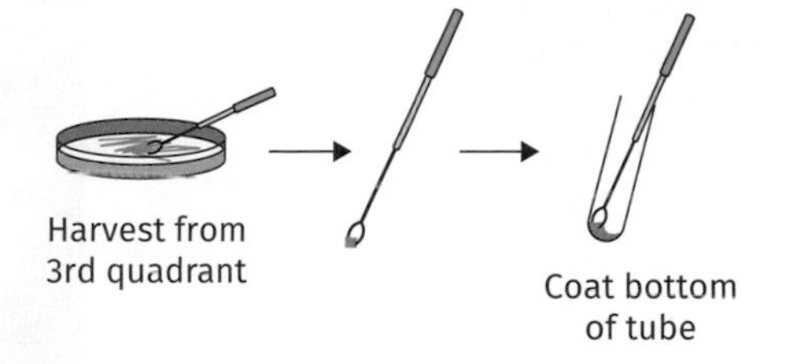

3. The bacterial cells are lysed (broken open) and the fatty acids from the cell membranes are saponified (converted to their sodium salts) by boiling for 30 min in sodium hydroxide and methanol.

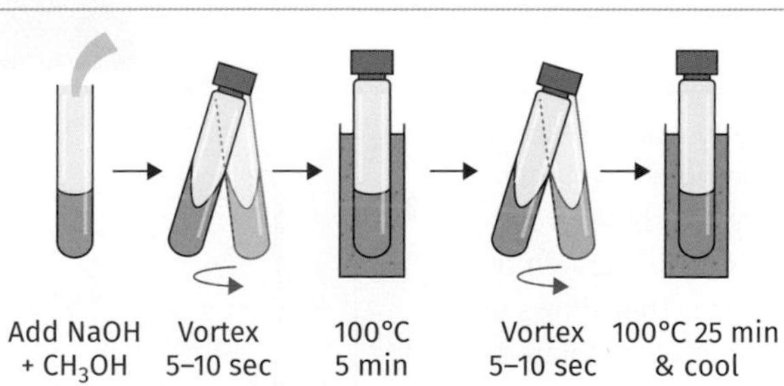

4. They are then methylated to fatty acid methyl esters (FAMEs) by heating to 80°C (+/− 1°C for 10 min (+/− 1 min) in hydrochloric acid and methanol.

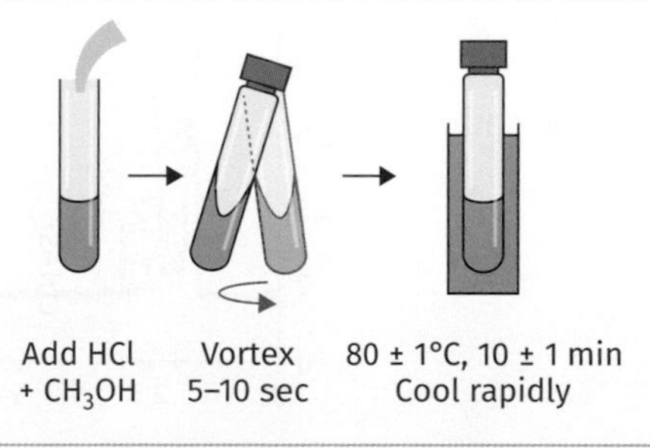

5. The FAMEs are then extracted from aqueous solution into an organic solvent containing equal parts hexane and methyl tert-butyl ether (MTBE). Mix for 10 min and remove aqueous (bottom phase).

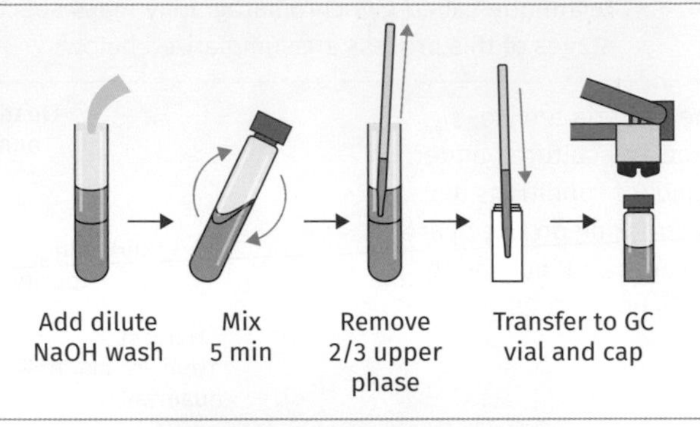

6. The FAMEs are washed in dilute sodium hydroxide solution for 5 min to remove free fatty acids and excess reagents in the aqueous (bottom) phase. Two-thirds of the organic (upper) phase is removed for GCMS analysis.

7. The FAMEs are separated by gas chromatography and the fatty acids present in the original sample are identified and quantified by mass spectrometry.

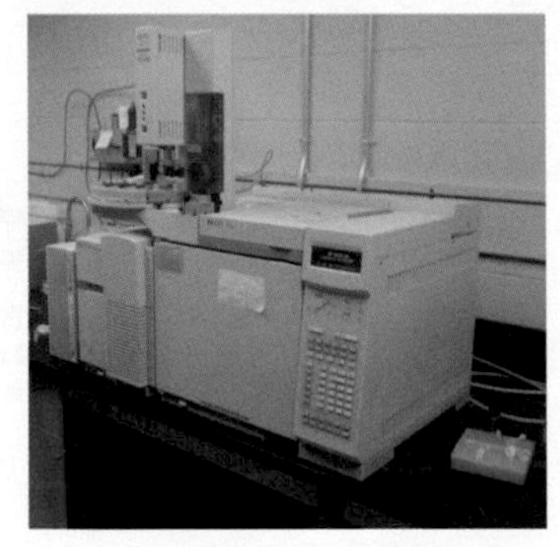

8. The fatty acid profiles are compared with a large library of profiles from reference bacteria in the database. This allows us to identify the isolated bacterium according to the positions and heights of each peak and the nearest fit in the database entries.

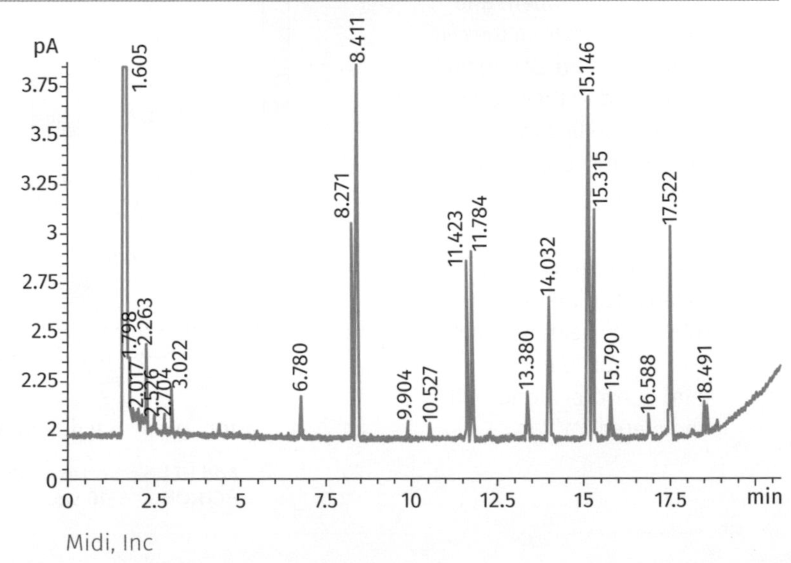

Midi, Inc

Certain phytopathogenic bacteria have also been classified directly according to their ability to cause disease on different host plants (that is, their pathogenicity). Sometimes, bacterial strains belonging to the same species differ in their ability to cause disease on different hosts. For this reason, the pathovar system of classification was developed whereby different pathovars are classified according to the defined range of host plants that each one affects. Many such pathovars have been established within different species of phytopathogenic bacteria, including many within the genera *Pseudomonas* and *Xanthomonas*. The identification of a particular pathovar then requires a series of pathogenicity tests whereby the pathogen is inoculated into a standard set of indicator host plants and the development of disease is monitored under defined growing conditions.

As was mentioned in Chapter 2, advances in molecular biology—and particularly methods for DNA analysis (bioinformatics)—offer rapid, reliable, and increasingly affordable opportunities for improved classification, identification, and detection of pathogens, including bacteria. Phenotypic comparisons based on physiological and pathological differences (as described above) can be time-consuming, laborious, and prone to high levels of experimental error. Comparisons of genotype, on the other hand, tend to be highly reproducible, being based on the cellular blueprint; the sequence of DNA nucleotides in the genome that ultimately defines each organism.

New methods for DNA amplification, fingerprinting, and sequencing continue to add to the plant pathologist's toolbox. These methods are similar to those that are used in forensic science to identify an unknown individual or to associate DNA found at a crime scene with a suspected criminal. Widely used DNA-based methods for classifying and identifying phytopathogenic bacteria include:

- Amplification of target DNA sequences—an automated process known as polymerase chain reaction (PCR) is used to amplify short DNA sequences of known length, which are unique to a specific pathogen. The amplified DNA (amplicon) is then recognized according to its size-related migration in agarose gel electrophoresis or its reaction with a fluorescent DNA probe, which hybridizes with the amplicon and simultaneously emits detectable light of known wavelength (real-time quantitative PCR).
- DNA fingerprinting—a variety of procedures are available, which nowadays mostly involve amplification of multiple recurring targets in the bacterial genome (sometimes called short tandem repeats (STRs) or microsatellites). This results in a series of amplicons which can be separated by electrophoresis and distinguished according to the variation in their sizes, giving a distinct profile for each species of bacteria.
- DNA barcoding—this involves the amplification of selected genes which are common to all bacteria, but which differ in DNA sequence

between species. The sequence of the amplified DNA is determined and compared with databases of sequences from fully identified reference bacteria. Sequences within the 16S ribosomal RNA gene are often used to establish the genus and species of a bacterium. More detailed comparisons may then involve sequence comparisons of multiple genes (often known as housekeeping genes) in procedures known as multi-locus sequence typing (MLST) or multilocus sequence analysis (MLSA).

- Comparative genomic sequencing—as the process for next-generation sequencing of whole bacterial genomes has become increasingly automated and more affordable and bioinformatics procedures to share and analyse large data sets of DNA sequences have become more available, the preferred method for classification of bacteria in many laboratories is now to compare whole genome sequences. This not only allows highly accurate identification and determination of their taxonomy, but also provides an insight into the differences between plant pathogenic bacteria in terms of their arsenal of enzymes, toxins, and other cellular products that contribute to their ability to infect and cause disease in different plants. We explore these genetic techniques further in Chapter 4.

The various genera of plant pathogenic bacteria and the main symptoms they cause are shown in Table 3.1.

How do phytopathogenic bacteria cause disease in plants?

Some bacterial plant pathogens are necrotrophs. This means the bacteria first kill the cells and then live on the contents of the dead cells within the still-living plant. Others, however, are biotrophic parasites—they live in and from living plant cells, acting as parasites. To cause disease, the bacteria have to be able to infect plant tissues, overcome any plant defence mechanisms, and induce symptoms. These symptoms, including the ones described in Table 3.1 and Figures 3.6–3.9, are the result of a number of possible pathogenicity factors associated with different bacteria and their interaction with the plant tissues.

For successful infection, bacteria have to get into the plant cells. They often use a wound as an entry point into the plant, although natural openings such as stomata, hydathodes, insect feeding or weather damage, and lateral root emergence can also provide entry for infections. Some bacteria, such as *Pseudomonas syringae*, can act as centres for ice formation on leaves and in so doing create frost damage through which they can infect (Figure 3.4). Many pathogenic bacteria secrete enzymes that help them to break down plant cell walls and membranes, making the tissues more accessible to infection, whilst liberating nutrients from the plant cells. Young plant tissues are usually more susceptible to infections because older tissues are more resistant to this kind of bacterial decomposition.

Figure 3.4 Leaves on a frosty morning can look very beautiful—yet some bacteria have evolved a way of using frost to damage the leaves and give themselves a way in!

Anthony Short

Just as we have defences against the entry of bacterial diseases into our bodies, so do plants. These plant defences to bacterial infections either involve the formation of corky or callous tissues around infection sites that are difficult for the bacteria to penetrate, or the production of metabolites that are toxic to the bacterial pathogens themselves, such as phytoalexins, polyphenoloxidases, gums, and phenolic compounds. Plants can also produce sticky proteins, such as cell wall lectins and other agglutinins that immobilize invading bacteria and prevent them from spreading. A key plant defence mechanism is the hypersensitive response (HR), whereby an interaction between a resistance (R) gene in the plant and an avirulence (*avr*) gene in the bacterium causes a very rapid and localized collapse of infected plant tissue which appears as a small necrotic spot. Biotrophic bacteria trapped in the necrotic tissue die and are unable to spread and cause further disease in the plant. As such, the plant sacrifices a small amount of tissue to save the whole plant.

Even if the bacterium finds a suitable entry point and manages to avoid the plant's defences, it still has to multiply and colonize the plant tissues before disease can develop and symptoms are observed. Successful infection is dependent on having a suitable environment which allows the bacteria to move into the tissues and to actively multiply. Most bacteria require free water for movement, a nutrient source (which usually comes from the damaged plant tissues), and a warm temperature suitable for multiplication. Hence, warm, humid conditions tend to favour development of most bacterial diseases. The triggering of disease often requires the size of a bacterial population to reach a critical threshold. At this point a build-up of certain biochemical signals stimulates the production of different pathogenicity factors, which in turn induce symptoms of disease in the host plant. This chain reaction is known as quorum sensing.

Figure 3.5 The leaf spots on this poinsettia are caused by an infection by *Xanthomonas axonopodis* pv. *poinsettiicola*

UK Crown Copyright—courtesy of Fera

There are four main types of pathogenicity factors produced by plant pathogenic bacteria, and each of them induces different types of disease symptoms:

- Toxins—These are often small glycoproteins and peptides. They can cause tissue chlorosis (yellowing, due to reaction with chlorophyll), necrotic white or brown leaf spots, or scab-like symptoms. They can also be transported around the plant's transpiration system and may therefore cause symptoms where they accumulate remotely from the point of infection. Bacterial leaf spots can be distinguished from those caused by other organisms by a chlorotic (yellow) halo caused by the toxin, as seen in Figure 3.5.

- Extracellular enzymes—These include pectinases, cellulases, proteases, lipases, and amylases. They break down plant tissues by degrading the middle lamella, which binds individual plant cells together, a process known as tissue maceration. This tissue breakdown leads to the onset of rotting symptoms, as shown in Figure 3.6.

- Extracellular polysaccharides—These are long chains of sugars that make up the coating of slime that covers many plant pathogenic bacteria, preventing them from drying out when outside of the host plant. When such bacteria invade the plant's xylem vessels the vessels become blocked with the slime, leading to water stress and resulting in leaf scorching and/or wilting symptoms (Figure 3.7).

- Growth hormones—Some phytopathogenic bacteria produce hormones such as indole acetic acid (IAA) and cytokinins, which stimulate abnormal plant growth, resulting in symptoms such as galls and tumours or excessive root formation—see Case study 3.2.

Figure 3.6 This potato tuber has rotted away as a result of infection by *Dickeya solani* infecting through the stolon from the mother plant

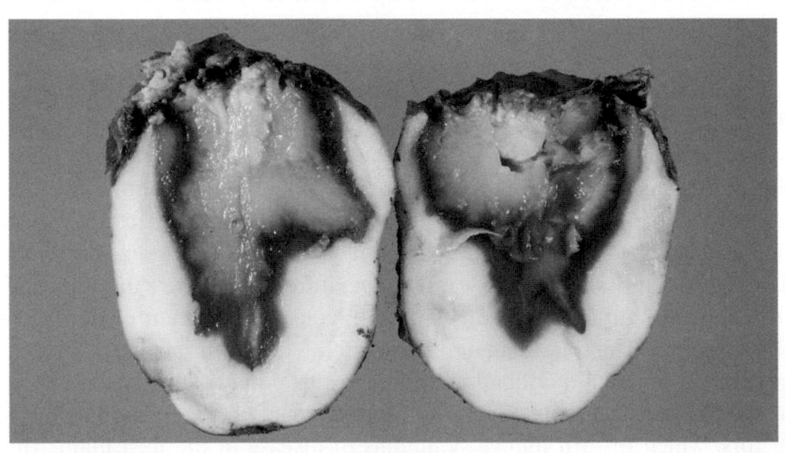

UK Crown Copyright—courtesy of Fera

Figure 3.7 The xylem vessels of this tomato are completely blocked by *Ralstonia solanacearum*—you can see this in the cut stem in the second image. Without water, the plant tissues wilt and die.

Case study 3.2

Crown gall and hairy root diseases caused by *Rhizobium* species

Crown gall and hairy root diseases are caused by soil-borne phytopathogenic bacteria belonging to the genus *Rhizobium* (formerly known as *Agrobacterium*). Pathogenic and non-pathogenic *Rhizobium* spp. occur worldwide in nurseries, orchards, and landscapes (cultivated and natural). They are **rhizosphere**-inhabiting bacteria and can be found on roots of host and non-host plants. Some members of the genus are beneficial bacteria, in that they form nodules on the roots of legumes (peas and beans) where they are protected and fix nitrogen into ammonium, which the plant uses as a food source. This is known as a mutualistic **symbiotic relationship**, beneficial to both bacteria and plant.

The plant pathogenic *Rhizobium* species carry pathogenicity genes on **plasmids**, which are short, circular DNA molecules (150–250 kb) that carry additional genes to those on the main bacterial chromosome and can replicate independently and be transmitted from one bacterium to another. Pathogenicity is not dependent on the *Rhizobium* species involved but whether the bacteria carries a tumor-inducing (Ti) or a root-inducing (Ri) plasmid, responsible for induction of crown gall and hairy root, respectively. *Rhizobium* species known to carry these plasmids are *R. radiobacter*, *R. rhizogenes*, *R. vitis*, and *R. rubi*.

Whereas pathogenic strains of *R. vitis* and *R. rubi* tend to specifically infect grapevine and *Rubus* species (e.g. raspberry and blackberry), pathogenic *R. radiobacter* and *R. rhizogenes* can collectively infect over 600 species of vegetables, fruits, ornamentals, weeds, and trees, and have a wider host range than any other plant pathogenic bacteria.

The infection process for these bacteria is unique. The pathogens, in soil or on infested plants, are disseminated by splashing rain, irrigation water, tools, wind, or insects, or may be introduced on already infected young plants or cuttings when used for propagation. The pathogen colonizes wounds made by pruning and cultivation, the natural emergence of lateral roots, frost injury, or insect and nematode feeding. It then attaches to living cells at the wound margin, and infects the plant by transferring part of its plasmid (known as the transfer-DNA or T-DNA) into the plant nuclear genome. Expression by the plant of genes on this transferred DNA results in excess hormone production, stimulating plant cell division and enlargement and gall or hairy root development. In addition, the T-DNA harbours genes that transform the host plant cells. The bacterial DNA directs the production and secretion of a variety of different compounds by the host plant cells, which are then used by the bacteria as a source of carbon and, in some instances, nitrogen.

The ability of these pathogenic *Rhizobium* species to transform plant cells has led to their widespread use in the process of genetic modification of plants—you will learn more about this in Chapter 6. In this case, tumour or root-inducing genes on the T-DNA are replaced by other target sequences (e.g. disease resistance genes) that can then be expressed in the transformed plant.

Plants infected with tumorigenic (Ti plasmid-carrying) strains usually develop galls (tumour-like swellings) on roots and stems below ground or at the root crown of the plant (the point at the base of a plant between the stem and the root), hence the name crown gall (Figure A(i)). On some hosts, galls can also occur on lateral roots and above ground on stems, canes, and vines. Crown gall disease is primarily a problem for nurserymen who grow woody plants and shrubs for landscapes and fruit production. Losses amount to millions of dollars annually from the destruction of diseased nursery trees.

Figure A (i) Crown galls caused by *Rhizobium* on tomatoes

(ii) Hairy root masses on cucumber plants, including the damage done to a commercial growing system

UK Crown Copyright—courtesy of Fera

Infection with rhizogenic (Ri plasmid-carrying) strains results in the excessive proliferation of lateral roots, resulting in less of the plant's energy being available for normal growth, which reduces the quality and yield of production. Financial losses due to hairy root (also known as root mat disease, illustrated in Figure A(ii)) in tomato and cucumber in parts of Northern Europe and the Russian Federation have resulted from increased costs of crop management (because the excessive roots block hydroponic irrigation systems), reduced yield and quality of fruit, and an increased susceptibility to root diseases.

Pause for thought

Which common human/animal disease does crown gall and hairy root remind you of and why? How do you think the ability to pass plasmids between bacterial cells and even between different species helps the pathogen?

How are bacterial plant diseases spread?

As described in Chapter 1, bacterial plant pathogens can spread over long distances within infected host plants or seeds (e.g. during their movement in trade) or locally between plants (e.g. from one plant to another within a crop or between neighbouring crops). They can also spread from active infections in diseased plants or from dormant or latent populations that may be associated with plants or seeds which do not cause disease symptoms. The bacteria may also survive in resting stages in the environment prior to coming into contact with their host plants and with environmental conditions that are suitable for the onset of infection and disease development. Some examples of plant diseases with details of how they are spread are shown in Figures 3.8 and 3.9.

Bacterial plant diseases can be spread in all of the common ways for pathogens to be dispersed, including soil-borne spread, wind-dispersed aerosols, water splash, by insect vectors, in contaminated irrigation water, or on contaminated tools and machinery.

One of the most effective ways to spread bacterial diseases is by moving infected seeds or other propagating material (young plants, cuttings, bulbs, tubers, corms, rhizomes, etc.) for growing in a new field, area, or country where the disease has not previously occurred. Since the bacteria already survive inside the infected host, the need for infection and avoidance of the plant defences is by-passed and the only requirement for disease development is then the onset of suitable warm and humid incubation conditions. It is very difficult to intercept such infected consignments and prevent them from being planted. This is especially the case if the incidence of infection is low or where infections are latent and symptomless, whereby they easily escape detection during inspections, sampling, and testing.

A number of bacterial plant diseases are seed-transmitted. In some cases, the bacteria are present as contamination on the seed surface. This contamination may be significantly reduced during procedures used to extract the seed from the fruit, or during dry seed storage conditions, and seed disinfection treatments can substantially reduce the risk of spread. However, some vascular pathogens can truly infect seeds internally, where they can remain protected and survive both disinfectant treatments and storage periods. The risk of disease development from infected seed depends on the rate of **seed transmission**, which in turn depends on the survival rate of the bacteria in the seed and the minimum population of the bacterium needed to infect the germinating seedling.

Figure 3.8 The Gram-negative bacterium *Ralstonia solanacearum*, which causes potato brown rot disease, can become established along watercourses by infecting alternative host plants of the perennial nightshade species *Solanum dulcamara* (commonly known as woody nightshade or bittersweet). Millions of bacteria can enter river water from the roots of infected plants and spread to potato crops downstream.

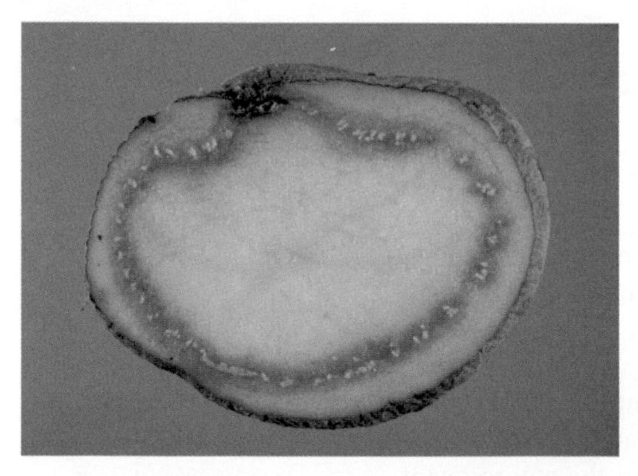

UK Crown Copyright—courtesy of Fera

Figure 3.9 The Gram-positive bacterium, *Clavibacter michiganensis* subsp. *sepedonicus*, which causes potato ring rot disease, can survive for long periods under cold, dry conditions on the surfaces of farm equipment or storage containers. Cleaning and disinfection is therefore a very important measure to make sure that the bacterium does not persist on the farm following any finding of this notifiable quarantine disease.

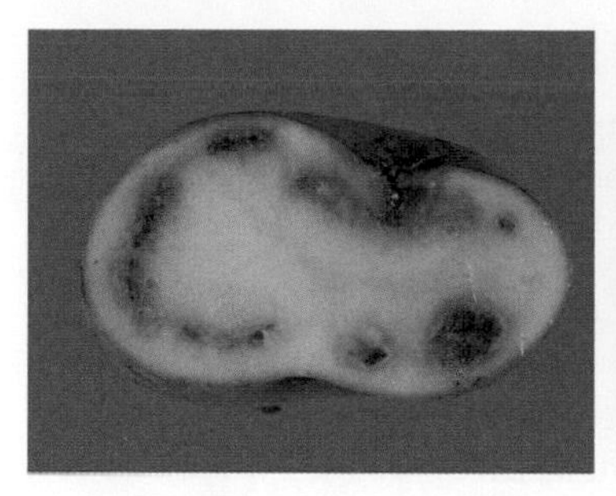

Chapter summary

- There are no approved chemical control measures and very few plant varieties with resistance to bacterial disease. Management therefore relies on successful early diagnosis and taking precautions to prevent introduction and spread of phytopathogenic bacteria in the first place.
- Only a small proportion of bacteria, belonging to three phyla (Gram-negative Proteobacteria, Gram-positive Actinobacteria, and Gram-positive Tenericutes), can cause disease of plants.
- Classification and identification of plant pathogenic bacteria has traditionally been a very laborious process but is benefiting from more rapid and accurate methods based on analyses of the DNA.
- Warm, humid conditions tend to favour development of most bacterial diseases.
- Plant pathogenic bacteria produce up to four types of pathogenicity factor—toxins, cell-wall degrading extracellular enzymes, extracellular polysaccharides, and/or growth hormones—which induce different types of symptoms.
- Long-distance spread of bacterial plant pathogens usually occurs during trade of infected propagation material (plants, cuttings, bulbs, seeds, etc.). Spread over shorter distances can occur as wind-dispersed aerosols, water splash, in insect vectors, in contaminated irrigation water, or on contaminated tools and machinery.
- Some bacterial pathogens can survive in soil for long periods between one susceptible crop and the next.

Further reading

Atlas of Plant Pathogenic Bacteria. http://www.atlasplantpathogenicbacteria.it

A database of images of field symptoms caused by plant pathogenic bacteria in different crops.

https://gd.eppo.int/

European and Mediterranean Plant Protection Organization (EPPO) Global database. Search for bacteria by name to find descriptions of the diseases they cause, their worldwide distribution, lists of their host plants, and images of the symptoms they cause.

http://www.bspp.org.uk/outreach/education.php

British Society for Plant Pathology. Search for 'bacteria' to find interesting factsheets and other resources describing bacterial plant diseases.

Janse, J.D. 2005. *Phytobacteriology: Principals and Practice*. CABI Publishing, CAB International, Wallingford, Oxfordshire, UK. 360 pp.

A comprehensive review of the scope, significance, impact, and control of bacterial plant disease. Includes a guide to the principles required for diagnosis of bacterial plant diseases and short descriptions of over 50 bacterial plant pathogens and the diseases they cause.

Schaad, N.W., Jones, J.B. and Chun, W. (eds) 2001. *Laboratory Guide for Identification of Plant Pathogenic Bacteria*, third edition. APS Press St Paul, MN, USA. 398 pp.

A laboratory manual describing traditional and molecular methods for the detection and identification of plant pathogenic bacteria.

Discussion questions

3.1 What main issues would you need to consider when planning a management strategy to avoid the introduction and spread of plant pathogenic bacteria onto a farm or nursery?

3.2 How do the different mechanisms by which bacteria infect plants contribute to the disease symptoms that they cause?

3.3 What are the main differences between the diseases described in Case studies 1 and 2 in terms of the ways in which they might be introduced, spread, and controlled?

4 VIRUSES AND VIRUS-LIKE PATHOGENS OF PLANTS

Adrian Fox

Plant viruses have evolved alongside their plant hosts over millions of years. The breadth of their diversity is potentially as great as the diversity of the plant species they infect. However, viruses are incredibly small, and they are obligate parasites: they cannot survive and reproduce outside of the cells of their plant hosts. As a result, they have presented many challenges to the scientists who study them. These challenges are only now being addressed with the advent of advanced **genomic sequencing techniques**. These new diagnostic technologies are also revolutionizing our understanding of the diversity and role of viruses in the wider environment.

Virus infections are not normally fatal to their host. They can, however, cause a range of symptoms that can reduce yield, prevent or reduce reproduction, or result in produce which simply cannot be sold, in the developed world at least, because of internal and external blemishes (Figure 4.1).

Viruses can infect terrestrial, aquatic, and marine plants. Some types of virus even play a vital role in regulating CO_2-fixing marine algae. As comparatively little is known about aquatic and marine species as pathogens, we will focus on viruses that attack plants on the land.

The chapter gives an overview of the history of plant virology, introduces the concepts of plant virus taxonomy and nomenclature, covers briefly the essential techniques used in diagnosing plant virus infections, and discusses the range of epidemiological factors which can drive plant virus outbreaks, focusing on the transmission of viruses. The chapter also presents cases studies focusing on important plant viruses from around the world.

Figure 4.1 Plant viruses can make plant leaves look amazing—but they cause untold damage to crops, to environments and economies

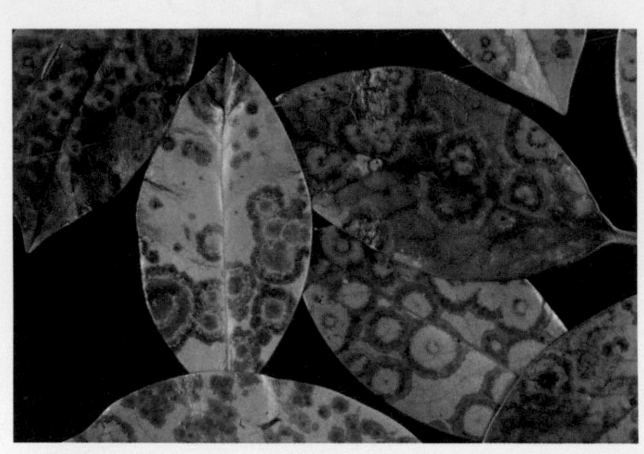

Sinclair Stammers/Science Photo Library

A brief introduction to plant virology

Viruses are very small particles, typically ranging in size between ten and a few hundred nanometres. They generally consist of a nucleic acid core surrounded by a protein coat. The nucleic acid carries the genetic information needed for translation (for production of proteins) and to allow more copies of the virus to be made through transcription and replication. To do this, the virus 'hijacks' the biochemical replication mechanisms of the infected host cell. Viruses cannot replicate outside of a host cell and are therefore obligate parasites.

The diversity of viruses is as great as the diversity of life on Earth: viruses have evolved to infect vertebrate animals (including people), invertebrates, algae, fungi, and bacteria, as well as vascular plants. While viruses that infect humans and other animals tend to receive a lot of media coverage, viruses of plants are not as widely known, yet plant viruses can have a devastating impact on both food production and the environment.

Unfortunately, it is difficult to estimate the overall losses due to viruses alone. One example we do have is for the State of Idaho, USA, where they estimated *Potato virus Y* to have caused a reduction in yield equivalent to over \$14 million (2010 costings). This figure could be increased to over \$33 million if the lost sales from 'value added' processing (frozen food products, prepared foods, crisps, etc.) and lost wages are then taken into account. That is a lot of money lost as a result of one virus in one state! Many viruses have the potential to affect food security, such as viruses of maize (*Zea mays*) and cassava (*Manihot esculenta*), both of which are subsistence crops in Africa and so are vital to the survival of millions of people.

As a biological discipline, plant virology is relatively young–only about 100 years old. However, it is widely thought that the first mention of plant viruses in literature comes in 752 CE, in a poem attributed to the Empress

Koken of Japan, writing about the unseasonally autumnal appearance of some beautiful shrubs known as grass roots (*Eupatorium* spp.):

> Perhaps it does frost
>
> In this village morn by morn
>
> For the grass I saw in the field of summertime
>
> Has already turned yellow.

This poem comprehensively describes one of the main characteristic symptoms of viral infections, where the plant turns yellow, also described as chlorosis. Although other pathogens can also cause this symptom, it is particularly linked to viral diseases.

Plant viruses have also contributed to the ruin of at least one major trading economy. In the Netherlands during the early seventeenth century, a craze developed for growing and collecting tulips. The most prized of these were rare tulips with a peculiar coloration, now known as colour breaking, which you can see in Figure 4.2. This coloration was caused by a virus, *Tulip breaking virus*. Trade in these tulips developed into a speculation market known as 'tulipomania'. The price of the tulips soared. One particularly rare tulip bulb sold for a price which was equivalent to buying (amongst other things) 70 tons of cereals, many barrels of wine and beer, over a dozen farm animals, a silver cup, and almost 500 kg weight of cheese!

The tulips became the subject of many fine art works, leading to the symptomatic flowers being known as 'Rembrandt tulips', named after the Dutch master painter. However, bulbs were bought months in advance of them showing signs of coloration and flowers exhibiting symptoms were both unpredictable and frail. The craze could not last. Demand for the tulips eased, supply increased, prices fell sharply, and many speculators lost vast amounts of money.

Figure 4.2 Tulip-breaking viruses create beautiful flowers but have led to financial ruin for many people in the past

Opachevsky Irina/Shutterstock.com

Plant virus research

Plant virus research started in earnest in the early twentieth century. As with medical research, the starting point for plant virus research has traditionally been the observation of symptoms and the investigation of acute diseases. Until recently, the vast majority of plant viruses discovered were first investigated due to the damage caused to food crops or because the symptoms of infection were particularly obvious (Figure 4.3). However, there are many plant diseases where the causal agent was thought to be a virus, but where no causal virus could be detected with the technology of the time. What's more, although most plant viruses are known due to them being plant pathogens, virus infections do not necessarily lead to symptoms in the host: some viruses can infect without symptoms and are not transmissible to other hosts.

The difficulties in detecting plant viruses led to plant virologists being at the forefront of plant pathology diagnostic development. They use the latest advanced molecular detection techniques, which in turn have led not just to an increase in the number of viruses known to be moving in global trade, but to an almost exponential increase in the discovery of previously uncharacterized viruses. There is a growing interest in these newly discovered viruses because scientists think they may have a number of positive uses as nanoparticles in a variety of areas of biology and medicine.

Taxonomy and classification

In this section, you will have a brief introduction to different types of virus groups and their general characteristics. Viruses are categorized by their morphology, their host range, the symptoms they cause, their genome type (RNA or DNA), and increasingly by their genetic sequence similarity.

The most noticeable thing about plant virology, in comparison to other biological disciplines, is that species are not named in Latin. Instead, plant

Figure 4.3 This electron micrograph shows *Beet necrotic yellow vein virus*, the causal agent of rhizomania disease of sugar beet

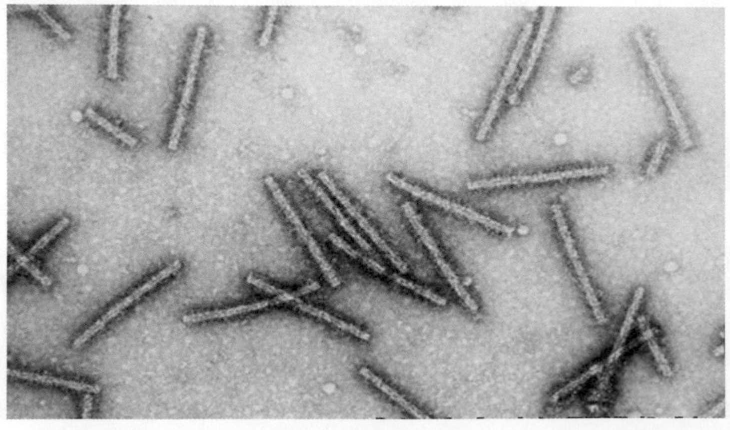

EPPO GD 2018

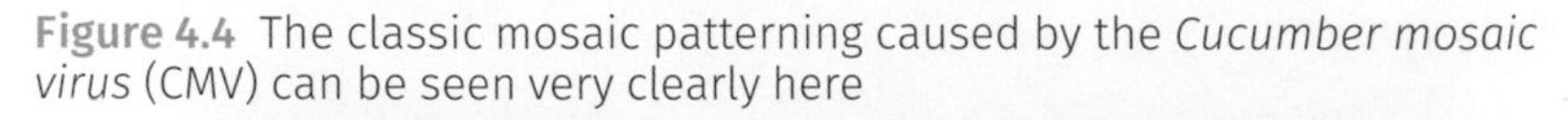

Figure 4.4 The classic mosaic patterning caused by the *Cucumber mosaic virus* (CMV) can be seen very clearly here

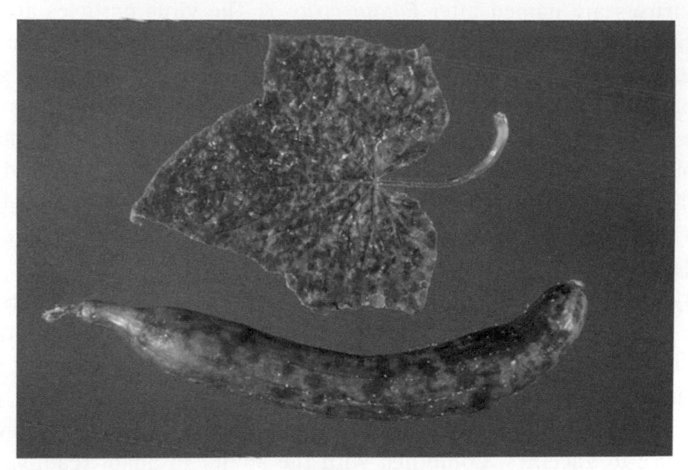

virus names use the following general approach: the common name of the plant species in which the virus was first discovered followed by the symptom in that host species. In other words, they adopt the format 'disease-virus'—for example, *Cucumber mosaic virus*. (CMV was first found on cucumber, where it induced irregular, yellow leaf spotting, known as mosaic patterning; you can see this in Figure 4.4.)

As you can see from this example, species names are written in English with the first letter of the species name capitalized and the rest of the name in lower case. A virus name is only italicized once the species (and its name) have been formally accepted by the International Committee on Taxonomy of Viruses (ICTV).

This system is ideal when a virus has a single known host and a limited range of symptomatic reactions are observed. However, CMV has one of the broadest-known host ranges of any plant virus, and has been reported to infect over 1200 species from at least 100 plant families covering both dicotyledonous and monocotyledonous plants. Across this range of hosts the virus induces a variety of symptoms including mosaics, ringspots, dwarfing/stunting, and 'fan-leaf'.

Where a virus cannot be demonstrated to be the causal agent of a disease symptom, the term 'associated' is used. When virus diseases were described this led to a large number of virus names being given, and it was only through meticulous biological, morphological, serological, and genomic characterization that these diseases have all been shown to be caused by the same pathogen.

Such complications are not even limited to different host species, but can also be observed on different cultivars of the same crop. Viruses are organized into distinct genera, usually named after the type of species. There are more than 60 genera of plant-infecting viruses and over 1400 accepted species, though many are unassigned to existing genera. In addition, another 2000 or so plant viruses have been described but are not yet accepted as species. Some examples of different virus types are summarized below.

Potyviruses

Potyviruses are named after *Potato virus Y*. The virus particles are filamentous (long and thin) and flexuous (curvy), typically around 20nm wide and 900nm long, with a genome size of 10 000 nucleotide bases. This group is probably the most successful virus genus on Earth due to their large range of hosts, their adaptability, and their ease of transmission, mainly by aphids. (See later in this chapter for details of ways in which viruses are transmitted from host to host.) Many of the viruses of this genus infect major food crops such as potato, tomato, and pepper, which has led to them being spread all over the world through trade in infected plants.

Potyviruses are single-stranded RNA viruses. RNA virus genomes replicate with high rates of mutation and consequently the population of a virus in an infected plant will be genetically diverse. This diversity gives the virus 'fitness' or the ability to readily adapt to evolutionary pressures. This is why many viruses which cause human diseases can develop resistance to vaccines. Combined with the ability to adapt readily, most potyviruses are transmitted by aphids (greenfly) in a non-persistent manner. (We explain the difference between persistent and non-persistent transmission by insects on page 92.) As a result, the action of an aphid picking up and passing on the virus is rapid, and when millions of aphids carrying infection are migrating simultaneously virus epidemics can occur quickly.

Many potyviruses are linked to high levels of economic damage: they reduce yield because infected plants produce less, or they impact the appearance of the plant products (for instance, causing fruit to be misshapen or marked) so the fruit cannot be sold. The two most prominent potyviruses are *Plum pox virus* (PPV) and *Potato virus Y* (PVY), both of which have a global distribution and cause many billions of pounds' worth of damage and lost production. PPV infects 'stone fruits' such as plums, cherries, peaches, and almonds, causing a loss of yield and fruit damage.

PVY has potentially the widest global distribution of any virus. It infects members of the potato family including potato, tomato, sweet peppers, and chillies, but the virus is also a serious problem for tobacco growers. The virus occurs anywhere potatoes are grown. Because the virus survives in living plant tissue, the virus can be passed on through vegetative reproduction of potato tubers. The virus has been spread all over the world with the movement of potato tubers ('seed potatoes'). Additionally, the spread of the aphid virus vectors is difficult to manage through the use of insecticides. PVY has evolved multiple strain types, which can give a range of symptoms in hosts—you can see an example in Figure 4.5.

PVY is also highly recombinant. This means that two different virus strains infecting a single host plant can exchange sections of nucleic acid and still function, creating a whole new strain of the virus which shares some characteristics of each of the original strains. Some of these recombinant strains give milder leaf symptoms, have greater fitness so they survive and reproduce more successfully, are more readily transmitted, and move more rapidly in infected plants than the origin strains. All of these factors combine to make PVY an incredibly difficult virus to manage. At the

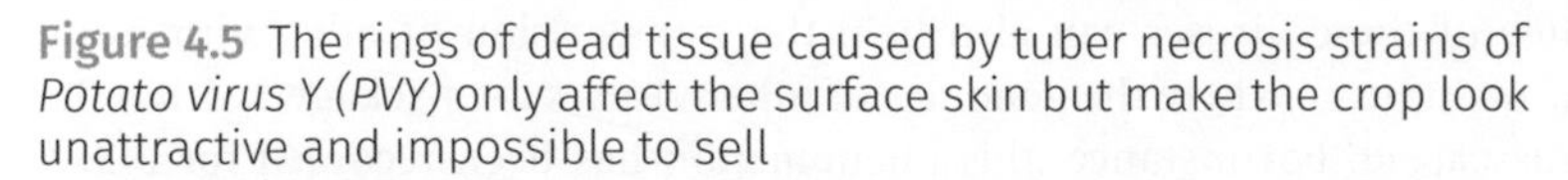

Figure 4.5 The rings of dead tissue caused by tuber necrosis strains of *Potato virus Y (PVY)* only affect the surface skin but make the crop look unattractive and impossible to sell

moment the virus is effectively managed in potato production through seed potato certification schemes. These schemes use visual inspection of seed potato crops to assess the levels of virus, and set acceptable thresholds for visual virus infection.

Luteoviruses

The Luteoviridae are a family of viruses also known as the 'yellow' viruses because many plants infected with these viruses become yellow in colour. (The name is derived from the Latin word for saffron yellow.) The virus particles are spherical and small, less than 30nm in diameter, with a genome size typically less than 6000 bases. Viruses from this family infect a broad range of crops, and are transmitted by aphids in a persistent, circulative manner (again, this is explained on pages 92–93). This means the virus takes a relatively long time to be acquired and circulate within the aphid before it can be transmitted into a new host. Some viruses in this family include:

- *Potato leafroll virus* (PLRV), which affects potatoes and causes stunting with leaves becoming thickened, rolled, and 'leathery', and can also lead to severe yield loss.
- *Barley yellow dwarf virus* (BYDV), which affects cereals, causing yellowing and stunting, and can cause yield losses up to 2.5 t/ha (over 30%) in winter-sown wheat.
- *Carrot red leaf virus* (CtRLV), which causes reddening of leaves in carrot crops and stunting, leading to yield losses.

Some of these viruses can also help the transmission of other viruses by **encapsidation**, where the genomes of other viruses are packaged within the virus capsid. For instance, this phenomenon has been recorded for PLRV, enabling aphid transmission of a secondary pathogen *Potato spindle tuber* viroid (PSTVd), which is not normally transmitted by aphids. A more commonly found example is CtRLV encapsidating *Carrot mottle virus* (CMoV) and Carrot red leaf virus associated RNA (CtRLVaRNA). These pathogens are collectively called the carrot motley dwarf complex (CMD).

Tospoviruses

Tospoviruses are named after the type member of the genus *Tomato spotted wilt virus* (TSWV). The particles of TSWV appear to be spherical under an electron microscope, and are approximately 90nm in diameter. The genome of tospoviruses is around 17 600 bases. Although tospoviruses affect plants, they are part of a family called the Bunyaviridae and are related to viruses such as *Hanta virus*, which causes haemorrhagic fever with kidney or lung failure in humans, and *Schmallenberg virus*, which causes fever, diarrhoea, and birth defects in cattle and sheep. Tospoviruses are transmitted by microscopic insects called **thrips** (Order Thysanoptera), illustrated in Figure 4.6. The main vector for TSWV is the western flower thrip, *Frankliniella occidentalis*. Other tospoviruses can be transmitted by different thrips, such as *Iris yellow spot virus* being transmitted by onion thrips (*Thrips tabaci*) in onion and leek crops. The viruses in this genus cause chlorotic and necrotic spotting on leaves, but can also cause stem necrosis.

TSWV is another broad host range virus, and can infect over 800 plant species. Because of the large range of host species and the

Figure 4.6 Thrips spread tospoviruses from one plant to another

Nigel Cattlin/Alamy Stock Photo

geographic spread of the vector, TSWV can be found all over the world. In warm-climate regions, such as the Mediterranean basin, the virus can cause severe damage to many economically important crops including potato, tomato, sweet and chilli peppers, peas, lettuce, papaya, etc; however, it can also cause damage to many ornamental (flower) crops such as chrysanthemums. In cooler regions, such as the UK, the virus tends to be a greater problem in greenhouse crops such as tomato and peppers, where growing conditions are ideal for the vector to survive all year round.

Given the close relationship to animal viruses it is thought that tospoviruses may have originated as insect viruses that developed into plant viruses. There is some evidence that thrips carrying TSWV show different feeding behaviour to virus-free insects: the insects with virus feed more often, and damage fewer cells near feeding sites, increasing the chance of successfully passing on the virus.

Nepoviruses

Nepoviruses are transmitted by nematodes, roundworms that live in the soil (see Figure 4.7) and can have serious economic impacts. The viruses are isometric (also known as polyhedral), around 33nm in diameter, and have a total genome length of around 16 000–17 000 bases. The name of the genus is a derivation of nematode-polyhedral virus.

Nepoviruses tend to have broad host ranges and affect many woody hosts, including fruit crops such as stone fruits (e.g. cherries), cane fruits (raspberries), soft fruits (strawberries), and some tree species, as well as ornamentals and many economically important crops. Nematodes carrying the virus can be unwittingly moved in soil attached to roots or in pots.

These viruses are also amongst the most readily transmitted through seed, and many are even transmitted through pollen. In some cropping systems

Figure 4.7 Microscope image of the nematode *Xiphinema americanum*. There may be half a million nematodes in a single teaspoon of soil—and many of them can carry plant viruses.

Jonathan D. Eisenback, Virginia Polytechnic Institute and State University, Bugwood.org

where plants are multiplied through cutting and grafting the virus can also be spread through infected cuttings. *Arabis mosaic virus* (ArMV) has been recorded to cause a wide range of symptoms across a large number of host plants, including yellowing and dwarfing of raspberry, yellow crinkle on strawberry, chlorotic stunt in cucumber and lettuce, ring and line patterns on ash trees, and yellow mosaic on rhubarb.

Other notable groups of viruses

- **Begomoviruses**, spread by whitefly, many of which cause major infections on staple crops such as cassava in developing countries, causing major food security problems.
- **Tobamoviruses**, which includes *Tobacco mosaic virus* (TMV). Primarily transmitted through plant-to-plant contact, these viruses are very robust as they have to survive outside the plant cells for long enough to be transmitted to their next host.

  TMV was one of the earliest viruses studied by virologists. However, the virus is still a major issue for the tobacco industry and there is evidence that many commercially available cigarettes contain tobacco with viable TMV. Not only does this virus survive the tobacco production process, but work carried out in the late 1960s found that sputum recovered from smokers with chronic lung disease also contained viable virus.

  TMV and other tobamoviruses also cause serious infections of tomatoes and peppers, and these viruses have been shown to remain viable in dry leaf debris for many months and even years, when other viruses would have become inactive.
- **Viroids**—a group of virus-like plant pathogens which exist without a protective virus coat, as naked RNA. Viroids can form robust structures which can survive temperatures up to 75°C and survive in dried leaf sap on skin for (at least) several hours, or on glasshouse windows and surfaces for many weeks. They are the cause of a range of plant diseases, including potato spindle tuber disease, tomato bunchy top disease, and chrysanthemum stunt.

Detection and diagnosis

All obligate pathogens—and so all plant viruses—present a diagnostic challenge because they cannot be isolated in culture. A consequence of this is that the discipline of plant virology has always been an early adopter of new diagnostic technologies, primarily those that can enable detection or identification of the pathogen directly from extracts of the infected host. The detection and diagnosis of plant viruses was originally carried out through symptom observation and transmission to uninfected plants, a method still in use in the laboratory today. However, other diagnostic techniques, such as demonstrating that the pathogenic agent was of small size by passing infective plant sap through a clay pot filter, have been superseded by a range of serological and molecular diagnostic methods. Even though plant virologists have a wealth of advanced techniques available to them, the starting point for virology diagnosis is still the observation of a plant showing signs of a virus-like symptom. It is important to remember, though, that plants under stress, such as those lacking nutrients or suffering herbicide damage, can also look as though they are showing virus symptoms.

Detection

Although many thousands of possible plant host–virus combinations are possible, the symptoms of viral infections can be generalized into a small number of categories, summarized in Table 4.1. These symptoms are the results of the sub-cellular impact of the virus infection that may result in changes to organelles such as chloroplasts, which would cause a leaf to lose colour, or even the denaturing of essential cell components resulting in cell death. Symptoms may be localized—that is, limited to certain parts of the plant structure—or systemic, whereby they occur throughout the plant.

One of the challenges in recognizing virus infections is the range and variability of symptom expression, which may be transient, can change in response to environmental conditions, or may even vary between cultivars (varieties) of the same crop. For example, infection with *Potato mop-top virus* was first described as causing shortened internodes of infected potato plants, giving the plant a dwarfed, bushy growth habit that looked a bit like an old-fashioned mop (see Figure 4.8a). However, this symptom is observed only in a limited range of potato varieties and the same virus is more commonly recognized by strong yellow, V-shaped markings on leaves (Figure 4.8b) and the presence of necrotic (dead tissue) arcs in potato tuber flesh (Figure 4.8c).

In many cases the more distinct symptoms can be a strong indication as to the genus or even the species of virus affecting a host plant. However, even with classical symptoms, diagnosis is made even more difficult because the precise nature of the symptoms shown can vary with the developmental stage of the plant at the time of infection.

Diagnostic techniques

Diagnostic technologies fall into two broad categories: non-targeted and targeted.

Non-targeted tests require no knowledge of the viruses you could expect to find in a sample. Until recently these methods either involved trying to infect host plants and waiting for symptoms to develop (see Scientific approach 4.1) or looking at direct sap preparations under an electron microscope for the presence of virus particles. More recently an advanced genetic sequencing technique (see next-generation sequencing in Scientific approach 4.1) has been developed, which allows the total genetic information in a sample to be analysed. This technique is leading to a new age of plant virus discovery.

Figure 4.8 The different effects of the *Potato mop-top virus* can be seen very clearly here, with the dwarfing of the plant leading to a bunchy appearance, yellow V-markings on the leaves, and internodal necrotic arcs known as 'spraing'

Muellek Josef/Shutterstock.com

Nigel Cattlin/Alamy Stock Photo

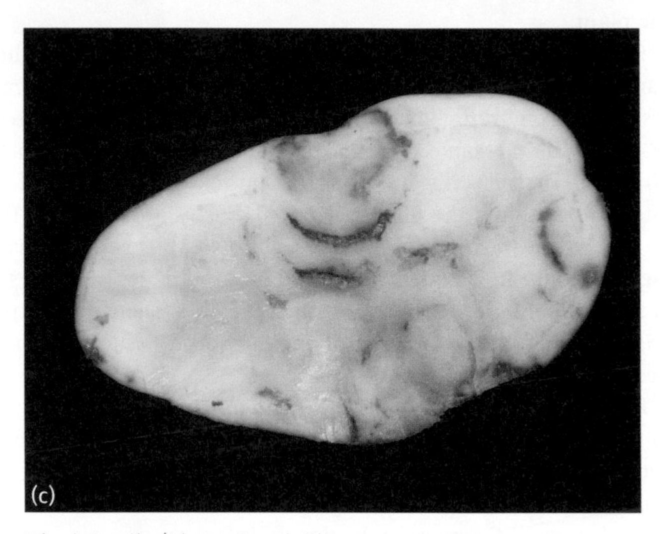

Nigel Cattlin/Alamy Stock Photo

Targeted technologies, such as enzyme-linked immunosorbent assay (ELISA) and polymerase chain reaction (PCR), are the mainstay of modern virology diagnostics. They are generally carried out in plates with multiple wells which allow 96, or even 384, test reactions to be carried out at the same time. This scale means that dozens of samples can be tested simultaneously for multiple viruses, giving a high throughput and quick turnaround of results.

However, targeted methods can be limited by their specificity. Using these methods requires a good knowledge of the viruses which are likely to be present in a species, otherwise you could end up doing hundreds of tests for every sample. The virus in the sample may be 'unknown' or it may be a genetically different strain, and a test may not be available for that virus.

Table 4.1 Common symptoms of viral diseases in plants

Symptom	Description	Example
Chlorosis, mosaics, and mottles	Form of yellowing of infected plant foliage called chlorosis. May be pale yellow colour all over, but in most cases appears as yellow spots, stripes, or a soft dappled appearance. Soft dapples through to small yellow spots: mottle. Larger, more distinct patches: mosaic (Figure 4.9).	**Figure 4.9** This fig plant infected with *fig mosaic virus* shows mosaic symptoms very clearly Image Courtesy Fera-Science Limited © Copyright Fera-Science
Ringspots, stripes, and line patterns	Yellow or discoloured patches may be distinct and form recognizable shapes and patterns. On long, thin leaves of monocotyledonous plants discoloration can follow the veins and the general leaf structure and form leaf stripes; yellow or necrotic rings known as ringspots can form, on the leaf, fruit, or tubers of infected plants (Figure 4.10).	**Figure 4.10** This rowan tree infected with *European mountain ash ringspot virus* shows the ringspots pattern symptoms very clearly  UK Crown Copyright—courtesy of Fera
Vein symptoms	Viruses move through a plant in the veins, and some viruses are limited to the vascular structures such that the cells in and around the veins may be the only plant part where symptoms are obvious. These symptoms include: *vein-banding*: the edges of veins look distinctly darker in appearance than surrounding leaf tissue; *vein-clearing*: veins lose their green colour and appear yellow, white, or silver (Figure 4.11); *necrosis*: the cells around the vascular tissue die.	**Figure 4.11** This sugar beet infected with *beet necrotic yellow vein virus* shows the vein chlorosis symptoms caused by this virus very clearly  Nigel Cattlin/Alamy Stock Photo

Table 4.1 Common symptoms of viral diseases in plants (*Cont.*)

Symptom	Description	Example
Deformation	Stunting or dwarfing is often the most obvious symptom of virus infection (Figure 4.12), largely due to reduction of chlorophyll in the leaves so the plant can't make food and therefore cannot grow properly. Some virus infections affect the appearance of the plant by affecting the growth rate of cells, speeding it up or slowing it down.	**Figure 4.12** The effects of *chrysanthemum stunt viroid* infection can clearly be seen in the infected plants compared to the healthy plant in the middle EPPO GD 2018
Necrosis	Viruses tend not to kill their hosts as they need them to survive, but in some cases cell death can occur in response to infection (Figure 4.13). This may be systemic or localized. In some plants there are specific genes responsible for this reaction.	**Figure 4.13** The brown marks on these tomatoes are necrosis as a result of infection with *Pepino mosaic virus*  UK Crown Copyright—courtesy of Fera

Scientific approach 4.1
Diagnostic techniques

1 Bioassay (sap inoculation, graft inoculation)

Because of the obligate nature of viruses, when working with live viruses plant virologists have to work directly with plants. 'Bioassay' is a broad term applied to this type of testing, which can be used in a number of ways.

The simplest form of bioassay is a 'grow-out' test, where the plants are grown on and checked for symptoms or tested using another method. Before biochemical and molecular diagnostics were available virus assessments were carried out by looking for symptoms in growing plants. Some seed testing, such as testing lettuce seed for *Lettuce mosaic virus*, is still carried out using this method because the test will show if the virus is both present and viable (causing infection) in the plants grown from seed. Additionally, the method can be used to **bioamplify** viruses which may be present at low concentrations, making them difficult to detect. This is the case with seed potato testing for PVY by ELISA, where the standard method is to take a sample of seed tubers and grow them for up to six weeks before testing for the presence of viruses by ELISA.

If plant viruses need to be maintained 'in culture', or the plant virologist is trying to demonstrate that an unknown pathogen is a virus, they use a technique called mechanical inoculation onto test plants. These plants are from species known to be susceptible to a wide range of viruses or they show distinct symptoms in response to infection with different virus groups. Such plants include several species from the tobacco family (*Nicotiana* spp.) and plants that are common field weeds such as *Chenopodium quinoa*.

An abrasive powder, a bit like fine sand, is dusted onto the leaf of test plants. Then a sap preparation is made by grinding a test sample in a buffer solution. This is gently rubbed onto the leaves of test plants, breaking the **epidermis** of the leaf. The cytoplasm from the damaged leaf will then stick to any virus particles in the sap preparation; as the leaf wound heals it will 'drag' the particles back inside the leaf, allowing virus infection to take hold. Plants are then grown on and observed regularly for up to four weeks where any symptoms are recorded, and further testing will be carried out. This approach can also be used to separate out viruses from mixed infections.

Some virus species, such as the poleroviruses, cannot be mechanically inoculated onto other host plants. However, these viruses can be transmitted to susceptible hosts using a method called grafting. A stem from an infected plant is cut and spliced onto a growing test plant and taped into position so that vascular tissue of the host and the sample can grow together. The virus can then be passed on to the host plant and can be checked for any symptoms that may develop.

2 Electron microscopy

A typical compound light microscope can usually give a maximum of 400× magnification, whereas the latest transmission electron microscopes can give a magnification of 5 000 000×, allowing us to image the protein structures on a virus capsid (see Figure A). This ability is very useful when it comes to identifying pathogens.

3 ELISA and serology

One of the most commonly used methods in plant virology is the ELISA method. This method uses **antisera** which have been produced to target specific proteins on the virus capsid. Because these proteins are different on each virus the antisera will only bind to the specific virus that they have been produced to detect.

Figure A High- and low-resolution electron micrographs can be used to help identify viruses like this Cowpea mosaic virus

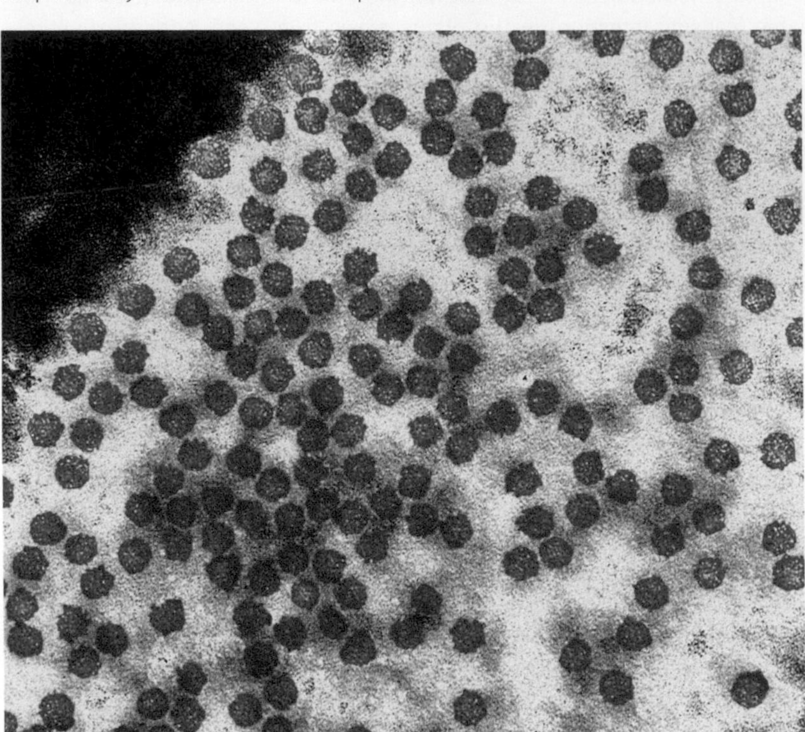

Centre for Bioimaging, Rothampsted Research/Science Photo Library

There are several different approaches to carrying out ELISA but most follow the same principles, as shown in overview in Figure B. Although the design and production of antibodies can be expensive and time consuming, once designed the antibodies can be produced relatively cheaply and in large quantities. In an automated process many samples can be tested at the same time, making ELISA testing cheap and robust. Commercial companies produce antisera, which can be purchased, for any of several hundred viruses, giving laboratories the ability to quickly and cheaply diagnose many common plant viruses. This technology has also been used to produce test kits for use at the point of infection, as you will see in Chapter 6.

Figure B A summary of the ELISA technique for identifying viruses

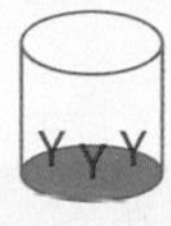

- Step 1—coating the plate with antisera.

- Step 2—adding plant sap and virus binding to antisera.

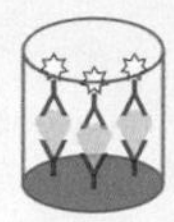

- Step 3—adding conjugate antisera, with a biochemical marker on the antibodies; this will bind to any virus that has bound to the coating antibodies.

- Step 4—addition of substrate, which reacts with the biochemical marker on the bound antibodies producing a colour change when virus is present which can be 'read' using a spectrophotometer.

ELISA does have some drawbacks, however:

- the method has a relative lack of sensitivity by comparison to molecular methods;
- some closely related viruses may cross-react with antisera for other viruses;
- some plant hosts can cause false-positive results by cross-reacting with antisera.

4 Molecular methods

Methods such as PCR and real-time PCR are also applied widely in plant virology. These methods have the advantage of being able to be rapidly designed for viruses once the genetic sequence of the virus is known. The key difference between these methods in virology is that classical PCR can only be used to detect DNA, so when testing is being carried out to detect RNA-based organisms the enzyme **reverse transcriptase** (RT) is used to make the RNA into DNA, which means the method can then be used to detect the RNA viruses. This reverse transcription may be done as a separate part of the method, or can be done as part of a single reaction step.

Sequence-based identification

Following PCR, the PCR products can be analysed at the genetic sequence level (a process called genome sequencing), giving an absolute identification

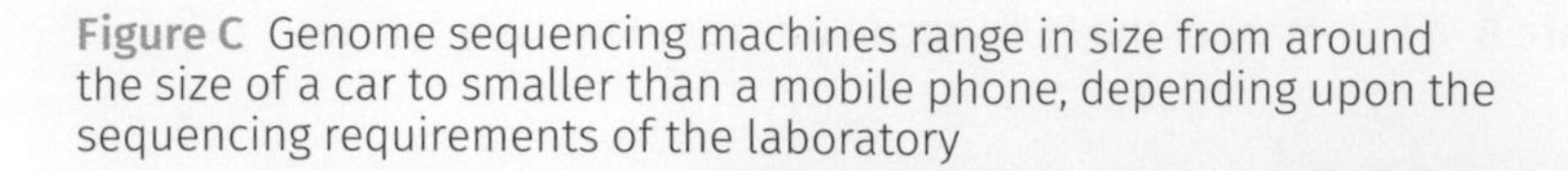

Figure C Genome sequencing machines range in size from around the size of a car to smaller than a mobile phone, depending upon the sequencing requirements of the laboratory

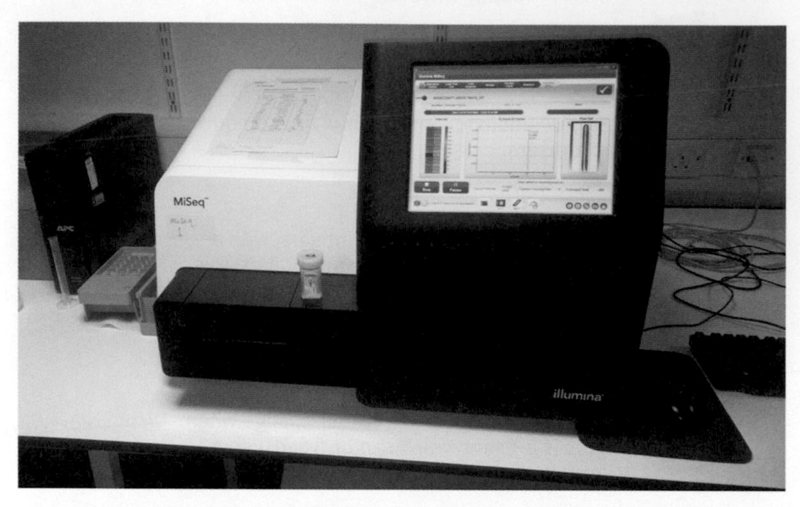

Dr I. M. Carr, University of Leeds

of the pathogen present in the sample. The conventional approach to sequencing is known as 'Sanger sequencing' after Frederick Sanger, the lead scientist who developed the method. Newer technologies, such as next-generation sequencing (NGS), allow genome sequencing to be carried out in a non-targeted way (Figure C).

Next-generation sequencing approaches allow the total genetic sequence of a sample to be analysed, and if used for viruses it can detect multiple viruses simultaneously. More crucially, unlike ELISA and PCR, this type of analysis requires no prior knowledge of what viruses can be expected in a sample, and it can detect viruses (and other pathogens) that were previously unknown. This approach has been used to identify the causal agent of carrot root necrosis, and has found several 'new' viruses in the same study on a carrot crop from a single field in Yorkshire.

Next-generation sequencing technology was used to identify the causal pathogens of a devastating maize (sweetcorn) disease in East Africa. Maize is both a commercial and a subsistence crop in Africa. In September 2011, a disease emerged in Kenya which was previously unknown in the region. The disease was unusual because plants would grow and look healthy for several weeks before they began to die; in some cases whole crops were dying (Figure D).

The disease was spreading rapidly and by April 2012 had been recorded in several regions of Kenya. Conventional ELISA diagnostics failed to give a conclusive diagnosis, so samples were analysed by NGS. This analysis suggested the presence of *Maize chlorotic mottle virus* (MCMV), a tymovirus, and *Sugarcane mosaic virus* (ScMV), a potyvirus. This combination of viruses had been previously recorded in the USA, causing a disease called Maize Lethal Necrosis, but had not been previously reported in Africa. The viruses found in the African samples were different to the previously reported viruses of

Figure D The new devastating disease of maize which has spread through Kenya and other African countries

CABI, 2018, Maise lethal necrosis disease [original text by Hannah Achieng Chore Odour]. In: Invasive Species Compendium. Wallingford, UK: CABI International www.cabi.org/isc

these species, and this is thought to be why they could not be detected by conventional ELISA diagnostics.

NGS has now been used to track the spread of the disease in Rwanda, Ethiopia, and Tanzania, where several different potyviruses have been found to cause the same symptom in co-infection with MCMV. This situation would make identifying the causal pathogen extremely difficult using targeted methods like ELISA and PCR.

The genetic sequence for the MCMV has also been used to design a real-time PCR test that is now being used to screen batches of maize seed to try and control the spread of this damaging disease.

Pause for thought

Why do we need a range of techniques to detect and diagnose plant diseases?

Transmission of viral plant diseases

We cannot cure viral human diseases, and we are no better at curing plant diseases caused by viruses. So a key part of understanding plant virus epidemiology, and being able to manage plant virus problems, is to understand how plant viruses move from host to host.

Because a plant virus is intrinsically linked to the cells of its host for replication and survival, a virus which infects a host systemically will also be found in the reproductive parts of the plant such as seeds and tuberous roots (e.g. potatoes). When these are planted, they can form a source of infection in previously uncontaminated areas. This type of movement 'in trade' is fundamental to the spread of many diseases, often over long distances, and has been the cause of multiple plant health outbreaks. However, when we consider transmission, we consider the movement of the virus from an infected host to an uninfected host.

Viruses cannot actively move between hosts, but need to be transmitted by a vector. A vector is another organism which can carry the virus and allow that virus to enter an uninfected host. Plant viruses can have a range of vectors, and some viruses even have multiple possible routes of transmission, enabling movement over both long and short distances.

Transmission of diseases can be spatial. Vectors can move infection into an area which had been free from a disease, or once infection has occurred in a crop it can be moved from host to host in a field. Within a field environment it is also important to remember that both inward and onward movement may be happening at the same time.

However, some vectors can move the virus into weed hosts between crop seasons, or the virus may be retained over periods where host plants are not available. This enables temporal transmission of viruses. The majority of virus vectors are insects such as aphids, whitefly, and thrips, but viruses may also be spread by nematodes and fungal plant pathogens, as well as mechanical transmission through cutting or plant-to-plant contact.

Sap-sucking insect transmission

In temperate regions, aphids are probably the most important vectors of viruses (see Figure 4.14), but some viruses are spread by whitefly, thrips, and other invertebrates such as mites. Species such as aphids are highly adapted plant parasites that can quickly build huge populations and migrate in large numbers. Transmission occurs through the act of feeding, and aphid behaviour can allow for rapid transmission of some viruses through probe feeding, where an aphid rapidly tries a plant to see if it is a suitable host and then moves on to another plant if not.

Different aphid-transmitted viruses will have one of three mechanisms of transmission:

1. Persistent transmission: The virus is taken up through feeding and passes through the gut into the haemolymph in the insect body, then circulates to the salivary glands, where it can be transmitted through feeding activity. The virus may replicate actively in the insect host. This can take some time, but once the insect has acquired the virus it will remain infectious through successive moults and can be infectious for life. Insecticides which attack the vectors can effectively reduce this type of viral transmission
2. Semi-persistent transmission: More rapid, but the virus will be lost from the vector more readily. Some viruses, such as the torradoviruses in their whitefly vectors, are stored in the foregut of the insect. This in effect makes the insect act like a flying syringe, acquiring the virus on feeding and then passing the virus on during subsequent feeds. It may take only a few minutes to acquire the virus via this mechanism, and the virus can be passed on in minutes, or at most a few hours.

Figure 4.14 Aphids penetrate the plant tissues and carry pathogens right into the transport system, to be carried all over the plant

Clouds Hill Imaging Ltd/Science Photo Library

3. Non-persistent transmission: Here the virus attaches directly onto the feeding mouthparts of the insect and so is transferred to the next plant very rapidly. For example, in aphids there are receptors on the tip of the aphid stylet, and the virus will bind to these through polypeptides. The potyviruses are transmitted by aphids in this way. Viruses with this mechanism can be transmitted by a relatively broad range of vector species.

For both non- and semi-persistently transmitted viruses the speed of transmission means that chemical management of the virus vectors through use of insecticides only has a very limited impact on spread of the viruses. The viruses are lost from the infectious vector when it moults.

Other vectors

Whilst insects are the main transmission route for viruses to get from plant to plant, there are a few other notable virus–vector relationships. These include:

- **Plant-parasitic nematode worms**: These microscopic, non-segmented worms pierce the roots of plants and feed on them. They can acquire and transmit viruses in a similar way to the sap-sucking insects. The two key groups of viruses transmitted in this way are the nepoviruses and the tobraviruses. The tobraviruses include *Tobacco rattle* *virus* (TRV), which has over 400 known host plant species. It is most commonly known for its effects in potato, where it can cause spraing (also called corky ringspot), similar to the tuber symptoms of *Potato mop-top virus*, that you saw in Figure 4.8. These nematodes can remain in the soil even when no potatoes are grown. The vector, and consequently the virus, survives by colonizing (and infecting) alternate host plants, including many field weeds. The next time potatoes are grown in the field the nematode vectors, and the viruses they carry, will return to affect the potato crop.

- **Fungal vectors:** Some fungal plant pathogens such as *Olpidium brassicae*, *O. virulentus*, *Polymyxa* spp., and *Spongospora subterranea* can be infected with plant pathogenic viruses, which they carry into the plants they infect. The *Polymyxa* fungi are largely pathogens of monocotyledonous plants such as cereals (e.g. wheat, barley, and oats) and rice. As a result, many of the viruses spread by these fungal vectors also affect cereals, or may affect crops grown as part of crop rotation, with crops such as sugar beet affected by *Beet necrotic yellow vein virus*, the cause of rhizomamia.
- The viruses often infect the fungal spores. These may live a very long time: the infected resting spores of some *Olpidium species* can harbour the virus for years. Fungal spores also allow the spread of viruses in huge numbers: a single zoospore from another of the *Olpidium* species can carry over 10 000 virus particles, so when the zoospore infects and penetrates the plant root, there is a strong likelihood the plant will be infected by the virus as well.

There are many other vectors not mentioned here: this is simply an overview of the range of vectors and mechanisms involved. Alongside the specialized vectors are many species that could be considered to be 'accidental vectors'. Contact-transmitted viruses and viroids can be transferred between infected plants and healthy plants on skin, clothing, hair, or animal fur. Any activity where plant sap comes into contact with a damaged plant could lead to virus transmission (e.g. tractor wheels, cattle movement, walking through a crop). Many viruses can also be transferred in the pollen of infected plants, which can be moved from plant to plant by bees pollinating flowers. This can then lead to seed-borne infections, which can transfer infections into subsequent crops. No wonder viral diseases are so widespread, so devastating, and so difficult to control.

Chapter summary

- Viruses range in size between ten and a few hundred nanometres, and consist of a nucleic acid core surrounded by a protein coat. They replicate by taking over the biochemical replication mechanisms of the infected host cell—they are obligate parasites.

- Although most plant viruses are known due to them being plant pathogens, virus infections do not necessarily lead to symptoms in the host. Some viruses can infect without symptoms and are not transmissible to other hosts. As such, plant viruses are now being investigated for their positive uses as nanoparticles.
- Unlike species from other fields of biology, viruses are not named in Latin. They are named after the host in which they are first detected, and the symptom observed on that host, such as *Cucumber mosaic virus*.
- Virus infections can lead to a range of symptoms. Often plants appear pale or yellow, or there may be changes to the way a plant grows or the shape of its leaves. In some cases, infection can lead to local dead patches on leaves or even plant death.
- Viruses cannot be seen directly and cannot be grown in artificial cultures, so virologists use a range of techniques to detect and diagnose the presence of virus infections. Often the best way to identify a virus is to analyse the genetic sequence of the virus.
- Viruses cannot move themselves from host to host and need vectors for transmission. These vectors may be insects, mites, nematode worms, or even fungal organisms. The main method that viruses move long distances is with the movement of plants in trade.

Further reading

Roossinck, M. J. 2016. *Virus: An Illustrated Guide to 101 Incredible Microbes*. Ivy Press.

Accessible book covering human, animal, and plant virology.

Boissoneault, L. 2017. *There Never Was a Real Tulip Fever*. Smithsonian.com https://www.smithsonianmag.com/history/there-never-was-real-tulip-fever-180964915/

Tulipomania background reading.

Descriptions of plant viruses. http://www.dpvweb.net/, Association of Applied Biologists.

Information on specific plant viruses.

BSPP New Disease Reports. https://www.ndrs.org.uk/

Up-to-date reports of outbreaks of plant viruses (and other plant pathogens).

http://www.bspp.org.uk/outreach/data%20sheets/TMV.pdf

BSPP outreach information on ***Tobacco mosaic virus.***

Discussion questions

4.1 Why are diagnostic technologies so important in plant virology?

4.2 Why don't viruses tend to kill their host?

4.3 Why can virus outbreaks be difficult to control?

5 PLANT BIOSECURITY

Dr Charles Lane

Introduction

The earlier chapters of this book highlight the variety of plant pathogens and plant diseases, as well as the devastation they can cause both to crops and to ecosystems. Although we tend to focus on plant diseases which affect food and ornamental crops, all of the tree diseases mentioned in this book can also have devastating impacts on whole ecosystems. Hundreds—even thousands—of species are dependent on any particular tree species. So when they are wiped out by disease, the whole ecosystem is undermined.

As international travel for people—and plants—has become easy and relatively cheap, the threat of plant diseases spreading across the globe has become more real and more pressing. What can we do to reduce these threats?

Biosecurity involves a set of precautions that aim to prevent the introduction and spread of harmful organisms. Although this is a relatively simple statement on the surface, once you start to delve deeper, establishing successful biosecurity raises a number of interesting and challenging issues. These are broadly associated with four key areas: trade pathways; international regulations; risk assessment; and introduction, spread, and detection of harmful organisms. The term 'biosecurity' can be applied to human, animal, and plant health, but the underlying principles, practices, and impacts are the same, whichever organisms are involved.

What is biosecurity?

Biosecurity means preventing the introduction and spread of harmful organisms, whether in plants, or in animals (including people). There are other definitions, but this one gives you a good working understanding of what is involved. While some of the basic principles are the same whatever organism is affected by the disease, others are very different, because of the differences in the life cycles and lifestyles of plants and animals.

Animal biosecurity

From a very early age we are taught the basic principles of biosecurity. Can you remember being told to wash your hands when you came in from playing before you were allowed to eat? This is one of the simplest and most effective ways to prevent the introduction and spread of harmful organisms such as the viruses which cause the common cold or life-threatening diseases such as influenza. We have all become increasingly familiar with what we need to do as a result of many national campaigns to encourage good biosecurity practice.

Good biosecurity practice is also now a routine part of a farmer's job, when raising livestock such as cows, pigs, and sheep or a variety of poultry including chickens, ducks, and turkeys. This good practice has been driven by outbreaks of harmful diseases such as foot and mouth disease (FMD; also known as *Aphthae epizooticae*) caused by a virus, bovine tuberculosis caused by a bacterium (*Mycobacterium bovis*), and avian or bird flu caused by viruses. Good biosecurity practices are important all the way through the livestock supply chain, from young animals to meat, to minimize the introduction and spread of pests and diseases that may not only be harmful to farm animals but also to humans.

Plant health

Biosecurity isn't just about farm animals. We are totally reliant on plants for the food that we eat, the oxygen in the air that we breathe, the land that we live on, and the places where we go to relax.

The introduction of harmful plant pests and diseases has both a direct and indirect impact on our health and well-being. The previous chapters have demonstrated the range of plant pathogens that can cause damage to plants, not only threatening our food security by reducing yield and quality (including making the food more perishable), but also damaging the environment we live in.

This chapter will go on and explore the principles and practices of good plant biosecurity.

Case study 5.1

No more chestnuts to roast? The spread of chestnut blight

Chestnut blight is a fungal plant disease caused by *Cryphonectria parasitica*—you can find out more about fungal plant diseases in Chapter 2. It has had a devastating impact on chestnut trees in many different countries, where it has been accidentally introduced from its origin in Asia (see Figure A).

For example, the pathogen was introduced into North America from Japan on infected nursery plants in 1904. It spread very rapidly in North America and by 1940 the majority of 3.5 billion American chestnut trees (*Castanea* spp.) were dead. This had a major impact economically, due to the value of the timber used in the construction of buildings and furniture, as well as the value of the sweet chestnut fruits harvested as a cash crop in the autumn. It also had a major environmental impact, as chestnut trees played a valuable role in the ecology of many areas: about 25% of the hardwood trees in the Appalachians were American chestnuts.

The fungus that causes chestnut blight was also introduced accidentally to southern Europe in 1938, possibly on infected plants from North America. The disease spread relatively slowly in Europe to begin with, but is now becoming more widespread. The UK remained disease free until 2011, when chestnut blight was found for the first time. The source of infection was traced back to planting material brought from France, European sweet chestnuts intended

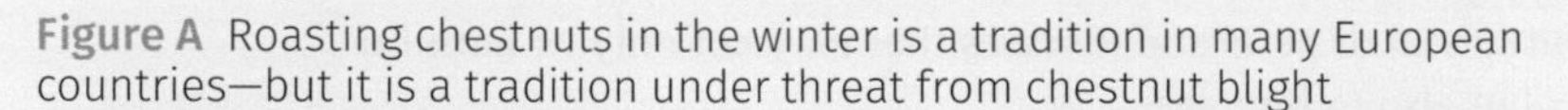

Figure A Roasting chestnuts in the winter is a tradition in many European countries—but it is a tradition under threat from chestnut blight

Karl Allgaeuer/Shutterstock.com

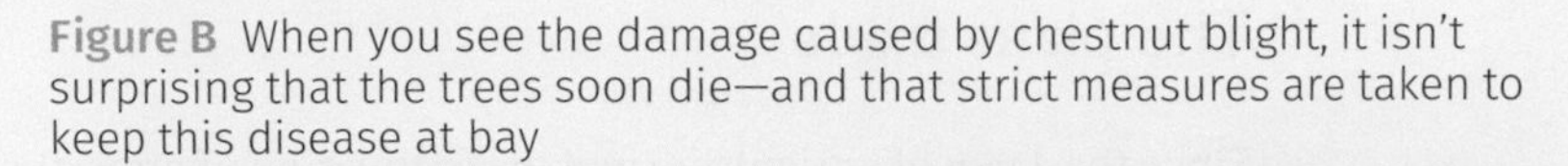

Figure B When you see the damage caused by chestnut blight, it isn't surprising that the trees soon die—and that strict measures are taken to keep this disease at bay

kevin/Shutterstock.com

for nut production. Swift action was taken, including uprooting and burning all of the infected sweet chestnut plants. Extensive surveys were established, up to 5 km radius from infected sites, to check that there were no signs of the disease.

The UK strengthened existing European Union (EU) plant health controls in response to these findings, including a requirement that host plants must originate from disease-free areas. The disease found in 2011 was successfully eradicated, but unfortunately chestnut blight was found in unrelated areas again in 2016 and 2017. Extensive efforts are now underway to prevent the spread of this very harmful disease in the UK (see Figure B). These measures include raising awareness about this damaging disease, putting in movement prohibitions on plant material that could harbour the pathogen (e.g. plants, logs or firewood, branches, and foliage) from known infected areas, and destroying infected material when it is found. Interestingly, the nuts we love to eat are not a major infection pathway. It is hoped these actions will prevent further introduction or spread of the chestnut blight pathogen within the UK.

? Pause for thought

What lessons have you learnt from the introduction of the chestnut blight pathogen and how might it be prevented in the future? Suggest reasons why it may never be possible to prevent the introduction of a new pathogen into the UK.

The plant trade

Since the dawn of agriculture and the cultivation of crops there has always been an opportunity for trade in plants and plant products. With time, the distances these commodities travel have lengthened from a few kilometres between villages to thousands of kilometres between continents, while the time taken between destinations has shortened to a matter of hours and days, as opposed to weeks and months. At the same time, more people have travelled further more often, experiencing new cultures and lifestyles. The demand to import exotic foodstuffs such as spices and herbs, tea and coffee has never ceased. Our expectations for year-round supply of common fruit and vegetables (such as strawberries and salad crops) but also exotic products (such as tropical fruits) grow year on year (Figure 5.1).

Currently, no country in the world is self-sufficient in terms of fulfilling its food needs, with many developing countries importing staple food crops such as cereals, rice, and maize; and developed countries supplementing domestic produce as part of their natural food security resilience. For example, the UK imports over 50% of its food. It also imports

Figure 5.1 Everyone expects to see exotic fruits on the shelves of supermarkets all year round—but the fruit we enjoy can carry diseases, not to people but to our native plants

Charles Lane, FSL

large volumes of wood and timber products for fuel, pulp for papermaking, and timber for construction, both from within the European Union and from North America and Asia. A large volume of ornamental plant material also enters the UK for gardens and urban woodland creation. The UK is also a successful exporter of plants and plant products—and no one wants diseased plants.

International trade: import and export

The movement of plants and plant products requires phytosanitary regulation to prevent the introduction of harmful plant pests and diseases. These controls are governed by international agreements (discussed in detail in the section 'How can good biosecurity reduce the threat?' on page 110), which put the onus on exporting countries to ensure these commodities are safe to trade. These agreements require the exporting country to certify that plants, plant products, and other regulated materials described have been inspected and/or tested as specified by the importing country, along with the issuing of a Phytosanitary Certificate. This document must accompany the goods and be checked on arrival. Therefore, there is a need for the importing country to be clear about what pests and diseases they wish to prohibit.

International regulations require these phytosanitary controls to be technically justified based on scientific evidence. This ensures that countries are not trying to prohibit trade in goods to protect their 'home market' from imports. This scientific evidence is gathered via a pest risk analysis, as explained in Scientific Approach 5.1.

It isn't just international travel that is controlled. Huge countries such as Australia and the USA have imposed regulations on the movement of plants and plant products between states as well.

Scientific approach 5.1
Pest risk analysis

Pest risk analysis (PRA) is composed of three elements: (1) risk assessment; (2) risk management; and (3) risk communication. We'll now consider each of these in turn.

Risk assessment

Risk assessment is a scientific process to identify and predict risk to health and life that may be associated with either a particular biosecurity hazard (referred to in plant health terms as a 'pest', which includes both pests and pathogens) or a specific commodity e.g. potato tubers, firewood, or strawberry plants). Risk assessment is reliant on scientific data that may be qualitative or quantitative. The decisions may not always be clear-cut (as there may be a lack of relevant information), resulting in some assumptions and subjectivity, which in turn lead to uncertainty in the risk.

Risk management

Risk management is concerned with ensuring an 'appropriate level of protection' (ALOP) for the level of risk to health or life determined in the risk assessment. It focuses on the consequences in terms of economic, environmental, or social impacts. The overall aim is to achieve the maximum risk reduction, while ensuring control measures are achievable and affordable, and that international trade agreements are met. A decision must be made as to whether any disease should be eradicated to prevent establishment, or contained to mitigate the impact upon the area affected and so prevent further spread, or whether it should be accepted that the pathogen cannot be controlled and will spread to its natural limit.

Risk communication

Pest risk analysis needs to take into account the views of stakeholders in a consultation process. These may include organizations such as farmers or foresters directly impacted by the introduction of a new pathogen, but also those indirectly affected in industries that are reliant on healthy plants for their business. These entities may include animal feed producers, hauliers, those involved with tourism, etc.

It is important for any communication to explain the consequences of the different management options proposed and the impacts on the stakeholders involved. For example, the PRA may conclude that the preferred option is to ban any further movement of a particular plant or plant product in order to prevent the introduction or spread of a pathogen. However, if your business was reliant on that particular plant then this would severely disrupt it.

Pause for thought

Pest risk assessment is described here as a scientific approach. How scientific do you think it is? Can you suggest ways in which the process might be more 'scientific'?

How are pathogens introduced and spread?

As you have seen in Chapters 2, 3, and 4, pathogens can spread in a variety of ways. Now we are going to pull all those strands together and consider all the ways the biosecurity of a country can be threatened, before looking at ways to overcome the threat.

Plant pathogens may be introduced into a new area through natural mechanisms such as air currents or watercourses (rivers or streams) or by human activities such as trade and travel. For each pathogen, there may be a single or multiple pathways of reaching new hosts–and those are dependent on both the characteristics of the pathogen and how the plants are traded and grown. For example, edible crops such as wheat and rice are raised from seeds, fruit crops such as strawberries are often sold as young plants, and garden shrubs and trees are sold either as young, bare-rooted plants or container-grown plants.

In developing countries, plants are often raised locally, and farmers regularly save seed from one year's crop to sow the next year. However, increasingly, plants used in places such as Europe and the USA are being grown in distant parts of the world. For example, a European plant breeder may develop a brand-new variety of a flowering plant. They will send this to another grower on a warmer continent (such as South America or Asia) to be multiplied up and where, in turn, seed is harvested or young plants are produced. Plants may then be sent to southern Europe to be grown on, before being transported by road or rail to be sold in northern Europe (Figure 5.2).

Figure 5.2 A new variety of flowering plant, although developed in Europe, will be sent to another grower on a warm continent such as South America or Asia to be multiplied up, and seed harvested or young plants produced. Plants may then be sent to southern Europe to be grown on before being transported by road or rail to be sold in Northern European stores.

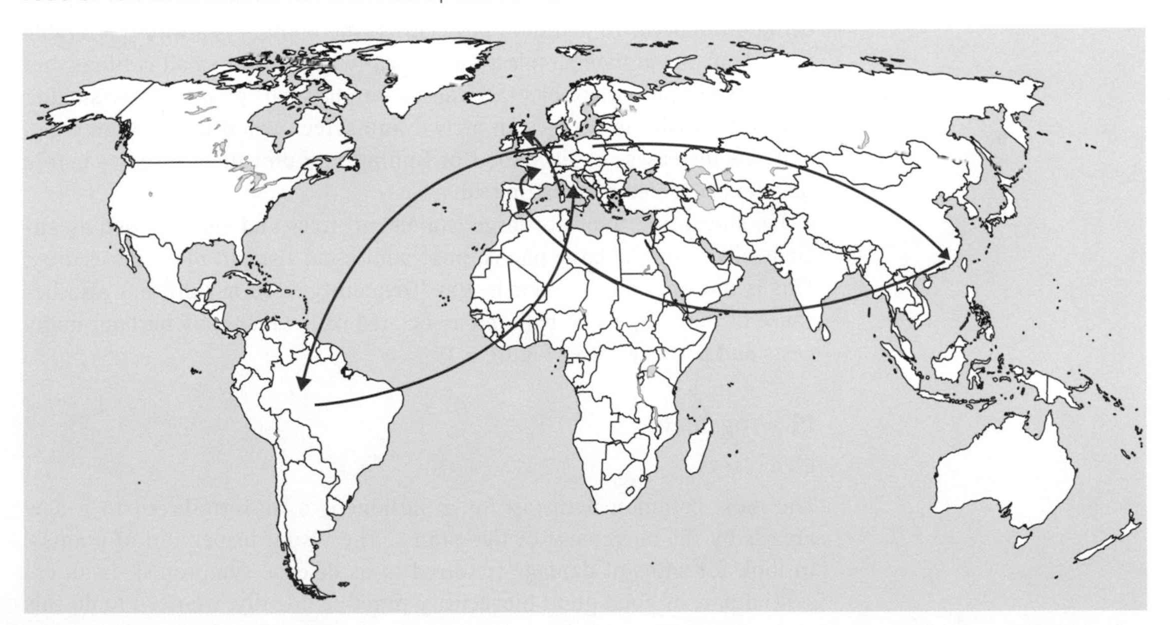

Figure 5.3 Using semi-mature and even mature trees at events such as weddings has become very popular—but these trees are often imported, and may bring in diseases which threaten the biosecurity of the country

Shutterstock

It is also important to remember that the larger the plant, the harder it is to inspect all over, and therefore, the greater the likelihood of a pathogen being introduced. To maintain biosecurity, therefore, it is always preferable to trade plants in their simplest form, ideally as seeds or small cuttings, but if this isn't appropriate or practical, as young small plants. These smaller plants are easier to inspect on arrival and screen when they are growing, and this increases the likelihood of finding any unwanted diseases before the plants are sold and distributed.

The increasing trend for large 'statement' trees and shrubs, such as ancient olive trees or large palms, pose additional risks to plant biosecurity. This is not only due to their height (frequently 5–10 m tall!) but also because of their large root ball and associated soil, which may harbour many pests and pathogens (see Figure 5.3).

Moving plants

Live plants

The most common pathway for a pathogen to be introduced to a new area is by the movement of live plants. The visual inspection of plants—to look for signs of damage (referred to as disease symptoms)—is an essential part of good plant biosecurity practice. Ideally, we need to do this before plants are shipped from one country to another, to prevent the movement of diseased plants. This visual examination can also be supported by laboratory testing of a representative sample of the consignment to be shipped, to give added confidence in freedom from plant pests and diseases.

However, as you have seen in previous chapters, many plant pathogens can live within plants without causing symptoms; this makes it difficult or impossible to determine which plants are infected. Therefore, plants inspected at the time of shipping may appear healthy, but may actually be infected. We can address this issue by testing the plant material again on arrival, or holding the plant material in a segregated area away from other growing plants—effectively keeping it in quarantine—and then inspecting it again before it can be sold on.

For example, a tree nursery that imports container-grown plants from southern Europe, and possibly further afield, will import plants during the dormant season (without leaves); by holding them for an entire growing season before selling them, they can check to make sure they are free of any pests and diseases. Certification schemes, where growers' crops (such as strawberry plants or seed potatoes) are inspected before harvest to determine disease presence and levels, are a common way to ensure farmers are buying good-quality seed and propagation material.

Parts of plants: fruits, vegetables, flowers

Fruits, vegetables, and flowers can also harbour plant pathogens. All of these are often grown in warmer climates, where not only is production cheaper, but also the climate allows them to be grown and supplied over a longer season to meet customer demands. So it is not uncommon for growers to buy in fruit, vegetables, and flowers from outside their own country and bring them to their nurseries to provide markets with a continual fresh supply (Figure 5.4). This movement has the potential to bring contaminated material into a plant production area, where these pathogens may escape and start causing damage, first to the grower's crops and then beyond.

Tubers, bulbs, and corms

The edible parts of plants and ornamental plants are frequently traded as live but dormant structures such as tubers (e.g. potatoes or cassava), bulbs (e.g. daffodils), and corms (e.g. begonias). Particular attention is paid to the trade in potato tubers both for planting and consumption as it is a worldwide staple food crop. Previous history, as witnessed by the devastating impact of *Phytophthora infestans* and the Irish potato famine in the 1850s (see Chapter 2), has demonstrated what can go wrong if good plant biosecurity measures are not in place and being followed.

Plant products

The movement of live plants as part of trade is not the only source of risk. Precautions also need to be in place for certain foods (referred to as plant products). Plant products include crops such as potatoes, grains (e.g. wheat, barley), and animal feedstuffs (e.g. peas and beans) that are sold for

Figure 5.4 We all love to have a variety of flowers to enjoy all year round—and many of them are grown far away in countries such as Kenya. This trade benefits everyone—but carries biosecurity risks which we must manage.

consumption, but which could accidentally be planted and used as seed for initiating new crops.

Plants are frequently grown for their edible products, such as tea (leaves of *Camellia sinesis*), rice, beans, flour, and tobacco. Any of these plant products can also act as a pathway for the spread of plant diseases. Fortunately, these materials are frequently processed prior to shipping (for example, the seed coat or husk is removed, or plant materials may be dried or milled); these processes usually remove or physically destroy contaminants and harmful pathogens. However, as you found out in Chapter 4, tobacco mosaic virus is known to survive the drying and curing carried out during the processing of raw tobacco leaves and can be transmitted on the hands of smokers to susceptible plants such as tobacco and tomato—so we always need to be vigilant!

Wood and wood packaging material (WPM)

High levels of plant biosecurity are needed concerning wood, wood products, and bark products. This is because of the large and diverse range of pests and pathogens which can inhabit both the bark and wood. Some of these organisms only live in the wood that is used for structural timber, fuel, or pulp for papermaking, whilst others only live in or on the bark. In order to reduce the risk of importing the pests and pathogens that live in the outer layers of the tree, wood can be debarked (referred to as 'squarewood') prior to movement.

The movement of 'roundwood' (where bark has not been removed) between countries is more risky and is therefore frequently prohibited. Campaigns to raise awareness about the risk of moving roundwood, especially from infected areas to clean areas, are also part of good biosecurity practice. Such campaigns are frequently aimed at campers, who may move small lots of firewood from their home to their campsites (which are frequently in wooded areas) and become unknowingly responsible for the introduction of a new pest or pathogen into an area previously free from it. In the USA, public awareness campaigns about the movement of firewood has been important in slowing down and preventing the introduction of emerald ash borer, a beautiful but devastatingly damaging pest of ash trees (see Figure 5.5).

The large-scale movement of low-grade wood for fuel (wood-burning power stations and home stoves or fireplaces) is increasing, but much of this material is pelleted and highly processed, which physically destroys pests and pathogens before shipping.

Wood products also pose a risk. These may include items such as packing cases, crates, and pallets, but may also include small pieces or scraps of wood (called dunnage), which are used to pack machinery or building materials that are frequently transported over large distances from other continents. This material should be free of bark and heat treated to prevent the introduction of pest and pathogens. Lower-quality trees, such as those

Figure 5.5 The emerald ash borer—this small beetle can destroy entire ash trees. Good biosecurity is needed to prevent its spread both within and between countries.

Herman Wong HM/Shutterstock.com

damaged by pests and pathogens, may be selectively thinned and the wood used for these wood products.

To help this process, international standards have been introduced. Wood packaging material and dunnage may be treated according to one of the measures in the International Standard for Phytosanitary Measures (ISPM 15). These include:

- debarked wood;
- heat treatment that achieves temperatures of 56°C for a minimum duration of 30 minutes throughout the entire consignment;
- microwaving material at 60°C for one minute;
- fumigation with methyl bromide.

All material being imported into the UK, for example, must display the appropriate mark to show it has been correctly treated.

Another potential pathway of introduction is the wood used in furniture manufacture. The wood must be bark-free and heat treated to meet ISPM 15, but furniture that is covered in material—such as chairs, sofas, and bed-heads—is difficult to inspect without causing damage. This means cheap furniture imported from abroad can still be a biosecurity hazard.

As you can see, we need to pay attention and inspect not just high-grade wood, but all wood-based products, if we are to maintain the biosecurity of our woodlands.

Animal vectors of disease

There are many different vectors of plant diseases—and for good biosecurity we need to be aware of them all and try to eliminate the risk that they will bring diseases to our plant populations. It isn't always easy!

Humans and their pets

People and their pets may unknowingly act as vectors for pathogens. This may be due to the accidental movement of contaminated plant parts (e.g. leaves and twigs) and soil on vehicles, footwear, paws, or clothing. However, simple wash-down procedures can greatly reduce the risks associated with this pathway by removing any contaminated soil or plant material. Disinfectants, such as chlorine or alcohol-based products, may be used as an additional precaution for footwear, but only after washing down to remove contaminated debris. Visitors to Australia have to sign a declaration that they are not bringing any soil into the country on their footwear. Be thoughtful about what you might be taking with you after walking your pet dog through woods or crops!

Wild animals

Plant pathogens may be transmitted from one place to another on the feet of birds, mammals, and other animals. When birds carry plant material from one place to another to build their nests, like the rook in Figure 5.6, they can transfer pathogens and pests at the same time.

Figure 5.6 Rooks, like many birds, move twigs and bark from one tree to another when they build their nests—a perhaps unexpected biosecurity hazard which it is almost impossible to prevent

Vishnevskiy Vasily/Shutterstock.com

Invertebrate pests

Insects which penetrate plant material to feed or to breed—for example, aphids and beetles (e.g. bark-boring beetles)—may spread pathogens. As you have seen in earlier chapters, aphids play an essential role in virus dispersal, while spittle bugs are involved in the dispersal of the bacterial pathogen *Xylella fastidiosa* that is currently killing thousands of olive trees in southern Europe, and beetles are a key part of the transmission of the fungus that causes Dutch elm disease, which has killed many of the elm trees in the UK.

Natural dispersal mechanisms

Air and water

As you have discovered in earlier chapters, plant pathogens produce reproductive structures, such as fungal spores, to aid dispersal. These may be released into air and water currents to aid local dispersal, but may also permit the introduction of that pathogen into a new area. Fungal diseases such as rusts, smuts, and mildews produce vast numbers of dry powdery spores, which are adapted for airborne dispersal. For example, American plant pathologists believe that soybean rust was introduced to North America from Colombia, South America during Hurricane Ivan in 2004.

Dispersal may also occur in freshwater courses such as streams and rivers which may cross regional and country boundaries leading to introduction to new areas. For example, the fungus-like organism *Phytophthora alni*, which infects the roots of alder trees, ultimately killing them, is spread in

Figure 5.7 The loss of alders as a result of *P. alni* infection has a major impact on the ecosystems involved

Anthony Short

freshwater systems. Infected trees develop black, weeping areas as you can see in Figure 5.7. Thick-walled survival spores (oospores) are produced in the roots, which in turn release motile zoospores that infect roots of healthy trees along river banks, leading to their death. It is therefore important to remember that natural mechanisms can also be responsible for the introduction and spread of pathogens.

Plant debris and soil

As you have discovered while looking at different types of plant pathogens, they are often adapted for survival during dormant periods such as cold winters or hot, dry summers by producing reproductive bodies found in plant debris and soil. The disposal of infected material may involve ploughing back into the ground, burning, or deep burial, but this may not always inactivate or kill all the pathogenic material. Burning and deep burial can also have negative environmental consequences.

Although composting is an effective way of breaking down unwanted plant material, it is not always effective at killing fungal pathogens. Effective composting requires temperatures to be maintained throughout the load at 55°C for 14 days or 65°C for seven days. This is rarely achieved by the composting done by gardeners in domestic situations, but commercial composting does maintain high temperatures throughout the entire load of material to kill pathogens, spores, etc.

How can good biosecurity reduce the threat?

It is one thing to recognize the many threats to the biosecurity of our plants, both the crops we rely on for food and the plants which make up our natural environment, but it is quite another thing to overcome them. For centuries, people have tried to control the spread of plant diseases, with varying degrees of success.

A brief history of quarantine and plant health control

The origins of quarantine go back to fourteenth-century Europe, when ships suspected of carrying the Black Death were prohibited from disembarking for about 40 days. As the Italian for 40 is 'quaranta', this led to word quarantine being used.

The first plant health measures aimed at controlling a plant disease were concerned with the destruction of the common barberry plant in Europe in the mid-seventeenth century. It was observed that, in its presence, cereal crops suffered greater loss due to stem rust. Amazingly, it wasn't until 200 years later that the basis of this control was understood: the fungus that causes stem rust, *Puccinia graminis*, needs the barberry plant to complete its life cycle and cause damage to cereals (see Chapter 2 for more details).

In the mid-nineteenth century, an American pest of vineyards was introduced into France (Figure 5.8) and soon spread not only through Europe,

Figure 5.8 The Great French Wine Blight, caused by an insect, was such a talking point of its day that it even generated a cartoon in the British satirical magazine *Punch* in 1890, showing Phylloxera attaching itself to and destroying the finest French wines

Public Domain/Wikimedia Commons

but throughout the rest of the world. In 1878, the International Convention on measures to be taken against Phylloxera was implemented and became the first international agreement designed to prevent the spread of a plant pest. Again, Chapter 2 gives you more details.

With the increasing globalization of trade in the late nineteenth and early twentieth centuries, further measures were taken against a range of plant pests and diseases including Colorado beetle, potato wart disease, and American gooseberry mildew.

In 1905, an international conference was held in Rome that led to the establishment of the International Institute of Agriculture (IIA). This evolved during the early twentieth century into the International Convention for the Protection of Plants in 1929. The period after the Second World War resulted in a lot of international collaboration, so in 1946, the Food and Agricultural Organization (FAO) succeeded the IIA; it is now the modern basis for all international plant health biosecurity, overseeing the **International Plant Protection Convention (IPPC)**.

As this short history demonstrates, people realized relatively early that regulations to prevent the introduction or spread of plant diseases need agreement at an international level. Luckily, in the twenty-first century, we have some largely effective plant disease controls in place. This is the result of both good international coordination and of having relatively few agreements and organizations involved in international phytosanitary activities; see Table 5.1.

These organizations are international. Nearer home, the UK has been part of the European Union (EU) for some time, although that is set to change. The EU has its own protections in place to regulate plant health across the region.

The EU plant health regime

The EU plant health regime applies a risk-based categorization of material from outside the EU. It classifies plant and plant products into one of these three groups:

- **Prohibited**: import to the EU is only permitted for specified purposes, such as research or trialling, under authority of a licence issued by a government.
- **Controlled**: requires a phytosanitary certificate issued by the plant protection services of the exporting country. There are EU requirements to undertake check inspections of this material at import.
- **Uncontrolled**: not subject to plant health controls and no notification is made on importation. This includes nearly all flower seeds, some cut flowers and fruit, most vegetables for consumption, biomass pallets, and furniture containing finished wood products.

A limited range of EU material which may harbour the most serious quarantine pests and diseases require a plant passport to facilitate their movement (see Figure 5.9). The plant passporting system also provides a framework of supplier information, so in the event of an outbreak of a new

Table 5.1 Major plant health organizations

Name	Initials	Logo	Role
World Trade Organization	WHO		The World Trade Organization General Agreement on Tariffs and Trade (WTO-GATT), which includes the agreement on the application of sanitary (humans and animals) and phytosanitary (plants) measures (WTO-SPS), which has been in place since the 1940s.
International Plant Protection Convention	IPPC	IPPC International Plant Protection Convention IPPC Secretariat	Founded by treaty in 1951, revised in 1997, and administered by a commission on phytosanitary measures under the United Nations Food and Agricultural Organization (FAO). This establishes the principle that all countries have a joint responsibility in plant quarantine to adhere to the rules agreed, without prejudice to obligations under other international agreements. Each country is required to establish an official National Plant Protection Organization (NPPO)—more about this in Chapter 6.

disease other potentially contaminated plants can be traced back and appropriate action taken.

Management options: prevention is better than cure

The best way to deal with plant diseases is to prevent them arriving in the first place—the purpose of the framework laid out by international regulations, standards, and collaboration. National plant protection organizations working collectively with the IPPC can all contribute to a global biosecurity continuum. Protective steps are taken at various stages.

Pre-border

National plant protection organizations work together to ensure effective application of control measures by producers and regulatory authorities,

Figure 5.9 Information on biosecurity and plant passports from the EU

European Commission Health and Food Safety

by, for example, enhancing training provision, knowledge, and expertise, as part of global initiatives to improve plant health. This helps make sure both importers and overseas exporters are aware of plant health biosecurity issues. These relationships lead to opportunities to share intelligence about new threats and have control measures in place to prevent their introduction.

Border

Import controls, including document checks and physical inspections, can reduce the likelihood of any pathogen getting into the country. It also checks the effectiveness of procedures in exporting countries which establish which of the three categories (Prohibited, Controlled, Uncontrolled) any plants or plant products should be in. Border controls also indicate where import regimes are not working effectively, so international actions can be taken. It also helps identify any additional surveillance work needed for the future and help understand global patterns of trade.

Post border

Action at the border alone can never reduce to zero the risk of known pathogens arriving. This means inland inspections at nursery sites and wider surveillance in urban and rural environments are essential (Figure 5.10). If an outbreak of a priority pathogen is identified, it is important to have a contingency plan in place to carry out a rapid and proportionate response. In some cases, eradication or containment will not be achievable, so the appropriate response will be for governments to work with the affected industry sectors to reduce impacts.

Figure 5.10 Inspectors regularly check plants at major growers and nurseries to prevent the spread of plant diseases around the country—and the world

Matylda Laurence/Shutterstock.com

In the case of an outbreak of a pathogen which is not a priority, there may be no role for government and it will be for industry and other stakeholders to decide how best to respond. To provide protection in the longer term, governments need to work with industry, non-governmental organizations (NGOs), landowners, and the public to increase resilience to the threats from pathogens.

Case study 5.2

Xylella fastidiosa—an international biosecurity challenge

Xylella fastidiosa is a bacterial plant pathogen (as discussed in Chapter 3), which is currently causing major plant health concerns in southern Europe and challenging plant biosecurity worldwide. *Xylella fastidiosa* causes Pierce's disease, which first became a problem in grapevines in Californian vineyards in the 1890s. When a vine becomes infected, the bacterium blocks the water-conducting vessels (xylem) leading to leaf scorch, wilt, and dieback; after one to five years the vine will die. The bacterium is spread by insect vectors: in the USA these are called sharpshooters, and in the UK 'spittlebugs' (Figure A).

The bacterium has four different subspecies that affect different plants and have separate origins. It has been recorded as damaging in excess of 300 different plant species including woody ornamentals, broadleaf trees, and herbaceous plants. The bacterium became of international concern

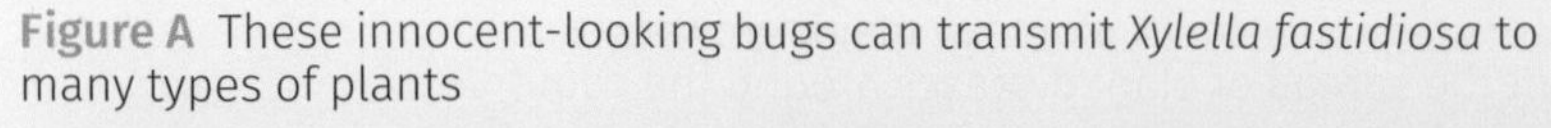

Figure A These innocent-looking bugs can transmit *Xylella fastidiosa* to many types of plants

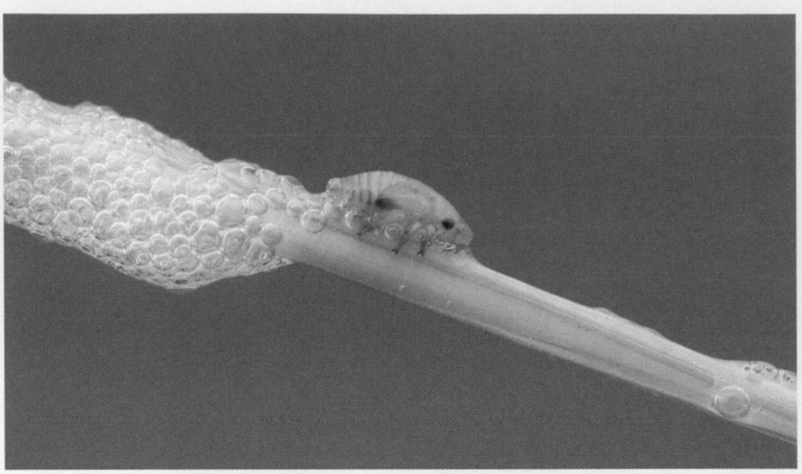

Sandra Standbridge/Shutterstock.com

when it was discovered in 2013 to have infected olive trees in southern Italy (Figure B); it has also spread to other, mainly southern, European countries. By 2015, up to 1 000 000 olive trees had been infected.

The bacterium is spread by native xylem-feeding spittlebugs which may only fly short distances (circa 100 m), but can also be carried over longer distances in the wind. Long-distance spread, and introduction into new areas, has occurred by the movement of infected plants.

In response to this emergency threat, *Xylella fastidiosa* has been subject to EU emergency phytosanitary measures based on a pest risk assessment and a stakeholder consultation process that identified it as a serious threat. Where the bacterium is present, measures are being taken to prevent further spread within the area. This includes surveillance to identify infected plants, establishment of 'infected' and 'buffer' zones (a 10-km band of land around the outer edge of the infected zone), and eradication of both infected trees and healthy potential hosts within a 100-m radius of the infected host tree(s). The buffer zone must be inspected twice yearly, and plants may not be moved out of the buffer zone into the non-infected zone, or from the infected zone into the buffer zone. Surveillance includes checking for infected plants and potential vectors. Additionally, control measures exist to keep the bacterium out of the UK and other unaffected countries. We may not have many olive trees in the UK, but *Xylella* could threaten our oak trees and other species.

To prevent further introduction, anyone importing host plants from the EU must ensure they are accompanied by a plant passport confirming they have been sourced from a disease-free area/site. If the bacterium is found, measures include the destruction of hosts within 100 m of any confirmed finding and a ban on movement from the area for high-risk specified host plants for five years. Alongside these measures, an active stakeholder and public awareness campaign and consultation is in place, to ensure everyone is aware of the risk and measures to prevent its introduction and spread.

Figure B Olive trees and the olives they produce are under threat—and biosecurity measures to protect them have swung into action

Both images; Anthony Short

A look to the future

In this chapter we have looked at the importance of human activities on plant biosecurity. As the world becomes 'smaller' due to even more rapid and frequent intercountry travel there is every expectation that the level of trade in plants and plant products will continue to grow. As we travel more and experience a wider diversity of food, plants, and habitats, our desire for non-homegrown plants and plant products is likely to increase. This globalization of trade and travel is also facilitated by greater digital connectivity around the world, providing further opportunity for the internationalization of markets.

This same globalization also has the opportunity to promote greater international collaboration and potentially the standardization of plant health regulations to facilitate the safe movement of plants and plant products.

The concept of 'prevention is better than cure' is a key theme, and working with countries around the world to prevent the export of infected plants and plant products in the first place is always the preferred option. The sharing of information will greatly help to give early warning of biosecurity threats, both in terms of the identification of new pathogens or commodities in trade.

As will be demonstrated in Chapter 6, the emergence of new diagnostic and surveillance technologies has the potential to permit earlier and more timely intervention, so increasing our chances of preventing the introduction of new diseases. However, plant pathogens may also be introduced by natural mechanisms such as air and water currents, so even with excellent pre-border and border controls there will always be a need for inland surveillance. Increasing awareness and understanding of plant biosecurity throughout the country in people of all ages and at all levels will greatly increase the number of 'eyes and ears' looking out for plant pests and diseases of concern. The evolution of **citizen science** to enhance government surveillance is an exciting prospect for the future.

Chapter summary

- Biosecurity involves a set of precautions that aim to prevent the introduction and spread of harmful organisms.
- Plant pathogens may be introduced into a new area through natural mechanisms or by human activities such as trade and travel.
- The most common pathway for a pathogen to be introduced to a new area is by the movement of live plants.
- The best way to prevent the introduction of plant diseases is to prevent them arriving in the first place.
- As plant trade has become international, plants diseases can spread further and faster than ever before, making international precautions more important.
- The movement of plants and plant products requires phytosanitary regulation to prevent the introduction of harmful plant pests and diseases.
- International regulations require phytosanitary controls to be technically justified based on scientific evidence gathered via a pest risk analysis.

Further reading

Protecting Plant Health: A Plant Biosecurity Strategy for Great Britain, April 2014

https://www.gov.uk/government/uploads/system/uploads/attachment_data/file/307355/pb14168-plant-health-strategy.pdf

UK government strategy concerning plant health and biosecurity.

OPAL tree health survey: https://www.opalexplorenature.org/treesurvey
Information about tree pests and pathogens and guidance on how to survey for them.

Observatree: http://www.observatree.org.uk/
Information about tree pests and pathogens and guidance on how to survey for them and the good of biosecurity practice.

Keep it Clean Campaign:
https://www.forestry.gov.uk/forestry/beeh-a6tek3
Advice and guidance on good biosecurity practice.

IPPC Food Security: https://www.ippc.int/en/themes/food-security/
Information about the activity of the FAO International Plant Protection Convention with respect to food security.

Discussion questions

5.1 Discuss some of the issues which arise because we like to eat non-seasonal fruit all year round. What are the ethical issues about this with respect to free movement of goods and risks of spreading plant diseases?

5.2 Who do you think should be responsible for maintaining the country's biosecurity? What role should government, stakeholders, and citizens play?

5.3 Research the terms 'biosecurity' and 'biosafety' and discuss the different definitions you find.

6 SURVEILLANCE: TRADITIONAL AND EMERGING TECHNIQUES USED BY PLANT HEALTH REGULATORS

Derek McCann

Why do **National Plant Protection Organizations** (NPPOs) around the world spend considerable time and money trying to find out if certain diseases are present or absent—and if they are found, tracking down how widespread they are?

The answer lies in gathering the evidence to support plant biosecurity, as described in Chapter 5, as NPPOs seek to prevent the damage that can be done to crop plants, garden plants, and the natural environment by plant pathogens ranging from viruses and bacteria to fungi (Figure 6.1).

The collecting and recording of information on plant disease is known as surveillance. It underpins:

- new laws, rules, or regulations designed to keep out an unwanted disease, which must be based on pest risk analysis
- the ability to export plants and plant produce to other countries, as this rests on an ability to show that diseases are not present
- the early detection of new and emerging diseases introduced despite phytosanitary controls being in place
- eradication programmes designed to get rid of an introduced disease, which need to know where to target actions and if those actions are successful
- an assessment of whether phytosanitary controls currently in place are successful.

Figure 6.1 Plant diseases may attack the plants that feed us or the plants which make up the environment we live in. Whatever the disease, NPPOs need to be on top of it, fast, and good surveillance makes this possible!

Cesare Palma/Shutterstock.com

It is obviously impossible to inspect every plant or tree that is grown commercially, is imported or exported, or is already present in the wider environment, so decisions about where to direct efforts need to be taken to gather as much accurate information as possible with the resources available.

What's more, we need to prevent phytosanitary controls being used to put unfair trade barriers in place, and to ensure other countries abide by international standards.

A clear definition of surveillance is therefore important, as it needs to be applied in the same way all over the world. The International Plant Protection Convention (IPPC) states:

- *Surveillance* is an official process which collects and records data on pest presence or absence by survey, monitoring, or other procedures.
- *Survey* is an official procedure conducted over a defined period of time to determine the characteristics of a pest population or to determine which species are present in an area.
- *Monitoring* is an official ongoing process to verify phytosanitary situations.

In this chapter you will discover how this surveillance is carried out—and the importance of getting it right!

General surveillance

General surveillance can be described as being passive. It doesn't involve teams of inspectors and scientists setting out to discover the presence of a specific disease. Instead it relies on the activities of others: it involves putting together information built up by them in the course of their work. Nevertheless, general surveillance is an important tool, used to alert NPPOs to the presence of a disease in their own or neighbouring countries and the countries they trade with.

This information can then be used to assess the importance of the disease, whether it is likely to spread or arrive, and the impact it will have. Ultimately, surveillance can help to determine whether specific surveys are needed.

The sources of information are many and vary in quality, accuracy, and availability.

Examples of sources used in general surveillance include:

- official reports by other countries. For example, Europhyt is the notification and rapid alert system run by the Directorate General for Health and Consumers of the European Commission;
- Regional Plant Protection Organizations (RPPO). For example, the European and Mediterranean Plant Protection Organization (EPPO) is an intergovernmental organization with 51 members, which is responsible for cooperation and harmonization in plant protection within the European and Mediterranean region;
- research papers published by universities and research institutes;
- scientific societies, amateur specialists, and citizen science;
- commercial and member diagnostic laboratories, e.g. The Royal Horticultural Society (RHS);
- landowners, farmers, foresters, wholesalers, retailers, advisors, and the general public.

Citizen science and surveillance

The discovery of the disease Chalara ash dieback in UK in 2012 (caused by the fungal pathogen *Hymenoscyphus fraxineus*) significantly raised public awareness about non-native plant pests and diseases. This outbreak led to a public outcry fuelled by reports in the press that up to 90% of all British ash trees could die. As ash is the third most common British tree, found frequently in hedgerows but also in urban areas, it is not surprising that people were very concerned. This triggered a Government review of plant health and biosecurity, which led to the publication in April 2014 of *Protecting Plant Health—A Plant Biosecurity Strategy for GB* by the Department for Environment, Food & Rural Affairs (Defra). This strategy recognized that:

> to provide the best protection, government, its agencies, industry, non-government organizations (NGOs), landowners, and the public require increased awareness of the risks to plants, together with the biosecurity measures to minimize them.

The report recognized the value of the input from the public and industry during the response to Chalara ash dieback, 'when they made a major contribution to the effort to identify its extent'. Increasing publicity to raise awareness with these groups is an essential part of enhancing plant biosecurity.

So how can the experts in plant disease raise awareness in the general public—and in the relevant industries—of the signs and symptoms of key plant diseases? There are many possible strategies. They include measures such as production of posters and publicity materials. The internet and social media can play a key role here too. 'Show gardens' at events attended by both public and industry are another valuable way of getting the message across.

The response to Chalara ash dieback clearly demonstrated the public's passion for our trees and concern about the impact of invasive pests and diseases on tree health. Asking members of the public to carry out surveillance activities as part of a larger initiative is referred to as citizen science.

The role of members of the public or amateur naturalists is a long-standing tradition in many countries. For example, Charles Darwin was an amateur naturalist who developed ideas about evolution. Today, many of the large-scale surveys that are recording the presence or absence of organisms such as birds, dolphins, ancient trees, ladybirds, butterflies, and moths are carried out by citizen scientists. However, these activities need to be managed carefully if data produced by citizen scientists are to be used to enhance government surveillance. Before launching a major national campaign to harness citizen science to survey tree health and look for damaging plant pests and diseases several key elements need to be in place:

- Awareness about plant pests and diseases and good biosecurity practice needs to be raised (Figure 6.2).
- Training resources need to be provided to help people identify host plants, the geographical position of the tree, and how to identify pests and diseases of concern.
- There need to be reliable and robust reporting mechanisms for collecting the data: mobile phones and digital cameras, with their global positioning system (GPS) and time functions, have an increasingly valuable role to play in this.

The power of citizen science has enormous potential in surveillance. By using interested and informed citizens, we can have thousands of people in every country on the lookout for the telltale signs of plant disease. The potential for improving plant health—if used wisely—is enormous.

The application of citizen science in the context of biosecurity has occurred in several countries with international collaboration. Projects such as the OPAL Tree Health Survey and Observatree in the UK, and the International Plant Sentinel Network and First Detectors in the USA, have all been successful in training citizen scientists to survey and record the presence of harmful plant pests and diseases. These projects have enhanced the work of government inspectors and play an important role as part of a country's early warning system for plant pests and diseases.

Figure 6.2 Events such as this Chalara ash dieback workshop help inform landowners, stakeholders, and the general public about plant diseases and what to look out for

Chalara ash dieback workshop

Grassington Town Hall, Grassington, Yorkshire

Thursday June 8th, 10am – 4pm

Do you want to know more about the recovery from ash dieback?

Do you know how to deal with ash dieback on your land?

This free workshop will bring together managers of ash research sites, concerned land-owners and managers of woodlands experiencing or threatened by Chalara ash dieback. The aim is to share information and experience and to renew partnerships in ash genetics and tree improvement research.

Speakers at the workshop will be:-

- **Dr Jo Clark** (Earth Trust) – The Future Trees Trust ash improvement programme and the Living Ash Project.
- **Ted Wilson** (Royal Forestry Society) – The biology of *Hymenoscyphus fraxineus*.
- **Dr Gabriel Hemery** (Sylva Foundation) – Getting people involved! The AshTag citizen science project
- **Ted Wilson** (Royal Forestry Society) - Silviculture and management of ash – Best practice advice for woodland managers.

After lunch, we will visit **Grass Woods**, a mature woodland owned by the Yorkshire Wildlife Trust which has been badly affected by Chalara ash dieback.

Numbers are limited, so to reserve your place at this important event, contact **Tim Rowland** at Future Trees Trust on **07896 834518** or e-mail him at Tim.Rowland@futuretrees.org

The **Living Ash Project** is a DEFRA-funded five-year project to identify resilient ash trees and to develop techniques to rapidly reproduce them. Learn more about the Living Ash Project at www.livingashproject.org.uk

This workshop is kindly supported by the Yorkshire Dales National Park. The **Living Ash Project** partners are:-

The Living Ash Project

Examples of citizen science projects include:

- **The Open Air Laboratories (OPAL) network:** This is a UK-wide citizen science initiative that allows anyone to get hands-on with nature, whatever their age, background, or level of ability. OPAL (Figure 6.3) runs a tree health survey which invites participants to monitor trees for telltale signs of key threats. It provides people with identification keys, so they know which type of tree is which, and gives them the symptoms to look out for.
- **Observatree:** This is an initiative between Government agencies and the Woodland Trust which uses trained volunteers to carry out inspections (Figure 6.4). Again they provide a toolkit of information and identification aids to get people looking out for new diseases including Siroccocus blight of cedars (first seen in the UK in 2014), Oriental chestnut gall wasp, new in 2015, and Xylella, which hasn't yet attacked trees in the UK.
- **Cape Citizen Science:** This initiative was set up in the Republic of South Africa. It is run by universities and research institutes, and encourages people across the country to become 'pathogen hunters' (Figure 6.5) and send in dying plants. The project is especially concerned with Phytophthoras.

Figure 6.3 A number of universities and organizations such as the Field Studies Council support OPAL and help train citizen scientists to spot plant diseases

OPAL (Open Air Laboratories)

Figure 6.4 Observatree is funded by a mixture of organizations such as the Department for Environment, Food & Rural Affairs (Defra) and the Woodland Trust and enables citizen scientists to be trained to recognize and report tree diseases

Observatree is a citizen science project led by Forest Reserach, in collaboration with key organisations. http://www.observatree.org.uk/

Figure 6.5 Poster for Cape Citizen Sciences

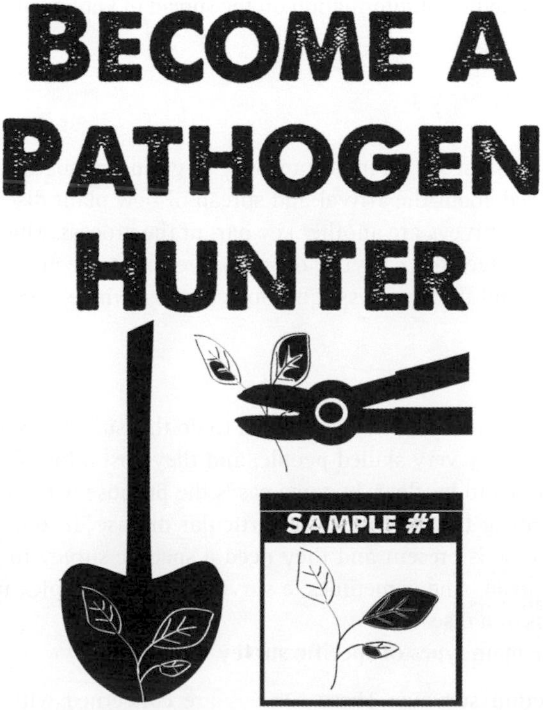

BECOME A PATHOGEN HUNTER

SAMPLE #1

Cape Citizen Science. http://citsci.co.za

Figure 6.6 The Forestry Commission provides a way for citizen scientists to confirm possible disease sightings through their Tree A!ert website

Reporting a disease

Governments in many countries provide easy ways for anyone to report suspicions and possible sightings. These include Australia's Exotic Plant Pest Hotline and New Zealand's Pest and Disease Hotline, which also include online forms.

An example in the UK is the Forestry Commission's Tree A!ert website (Figure 6.6).

The site allows anyone to send in a photograph and flag the location of a tree they think might be diseased. This will then be looked at by an expert. If the symptoms look like a new or regulated disease an inspector can be sent out to investigate further. Even if the report is not a regulated disease, it provides an invaluable source of information on the spread of known endemic diseases.

Specific surveys

General surveillance is an important way of giving regulators the information they need about the arrival and spread of new plant diseases, but specific, targeted surveys are another key part of the process. Much of the work of NPPOs involves the carrying out of surveys, and each one needs to be designed around the four basic questions of why, where, when, and how?

Why?

This question is really 'Why do we need to do this survey?' Specific surveys are carried out by very skilled people, and they cost a lot of money, so the objective needs to be clear. In some cases the purpose is to detect the presence or confirm the absence of a particular disease. In others, regulators know a disease is present and they need a specific survey to find out how far it has spread. And sometimes, a survey is needed to pick up changes in the progress of a disease.

The three main types of specific survey are:

- **Detection surveys**: These surveys are concerned with the presence or absence of a regulated disease (that is, to show a disease has appeared), but are not designed to provide information on outbreak size.

For an NPPO the aim is to find such diseases at a stage where the number and size of each find is small enough for them to be eradicated effectively. NPPOs often target their detection surveys to areas and locations where it is more likely to find disease introductions—for example, at growing sites where lots of plants are imported.

- **Delimiting surveys:** One of the first responses following the detection of a regulated disease is to carry out a delimiting survey, which is designed to establish the boundaries of an outbreak area. This information is vital to making decisions on whether eradication action is feasible—and cost effective—as well as defining the areas where action needs to be taken.
- **Monitoring surveys:** Designed to detect changes in a known disease outbreak. These are used to record changes in the geographic spread and population levels affected by a disease over time. This information is used to evaluate eradication action and ultimately whether it has been successful.

Where and when?

Inspecting every farm, forest, park, garden, nursery, or shop that may be vulnerable to a regulated disease is beyond the resources of any NPPO. Therefore the best they can do is to target their efforts to places where new and regulated diseases are most likely to be detected. This targeting can either be by choosing the location to inspect at, the time an inspection is undertaken, or more usually, a combination of both.

There are many factors that determine which sites are chosen for survey. These include:

- Geography: There are areas where lots of people grow particular plants to sell. This can be due to favourable growing conditions—for example, soil types and access to irrigation. Other factors are commercial: good access to transport networks and the closeness of a market. The clustering of places growing similar plants means there is a higher risk of disease introduction and spread.
- Size: The more grown, the more there is to inspect in one visit.
- Source: Origin of input material. For example, ornamental cuttings and seed from other countries may carry a higher risk than those that are home-grown.
- Storage: Harvested products are often stored at central locations, which makes them relatively easy to sample.
- Distribution: Many plants in trade are routed through central distribution depots. Again, these are good places to carry out a survey.
- History: Previous disease reports in an area mean inspectors are more likely to return.
- Host plants: It is only worth surveying where host plants that can become infected are present.
- Climate: Diseases, host plants, and disease vectors have optimal growing conditions, for example average temperature and rainfall.

Figure 6.7 Time of year and the stage of growth of a plant all affect whether we can detect potentially deadly diseases in plants from trees to food crops. For example, the twigs of a horse chestnut in spring (a) and in full summer leaf (b) look incredibly different, whether infected by pathogens or not.

Anthony Short

Alamy

The timing of any survey is influenced by a number of factors. One of these is the life cycle of the disease. For example, late potato blight prefers high moisture conditions and moderate temperatures for spore production, so it is more likely to occur at the end of the growing season in the UK. If a plant disease is spread by a vector, the life cycle of the vector also has to be taken into account. For example, as you saw in Chapter 5, the meadow spittle bug transmits *Xylella fastidiosa*. The pathogen is maintained in the gut of the vector and adults need to feed on infected plants in order to acquire and transmit the pathogen. The vectors are not active in the winter months so it is important to know the life cycle of the pest and plan surveys accordingly.

Nature's calendar also needs to be taken into account (Figure 6.7). Different plant diseases affect different parts of the plant and so the timing of budburst, first flower, fruiting, leaf fall, etc. in the host plant has to be taken into account when considering when to survey.

And finally, surveyors need to consider when the symptoms will appear. There is often a long delay between infection of a plant and the appearance of symptoms, especially in trees. Inspectors need to know where to look (e.g. in active growth or in harvested crop) and exactly what they are looking for.

How?

Having decided on the what, where, and when, the success of any survey—the ability to actually find the disease you are looking for—rests on how it is carried out.

Several basic criteria must be considered when you are planning any survey. These include sample size, sample unit, tolerances, detection levels, confidence level, and efficacy of detection. What exactly do all of these terms mean?

- **Confidence level**: Following an inspection, how confident can the surveyor be in their results? The level of accuracy that will be demanded from any survey needs to be decided by any NPPO before it is carried out. If it is set too high, it becomes difficult to achieve, but if the bar is set too low, the survey has little value. Consequently, NPPOs often work to a 95% confidence level. This means that, on average, the inspection result is correct 95% of the time. However, the converse is also true: the result of 5% of inspections will be wrong. This has to be taken into consideration in any decisions taken as a result of the findings of the survey.

- **Detection level:** The detection level is the smallest percentage of disease presence that can be detected, given the survey's sampling method and efficacy of detection, and the level of confidence the NPPO chooses to work at. A decision on what detection level needs to be set can be made by the NPPO before the survey begins. The decision will be in part based on knowledge of actions taken elsewhere to ensure disease is not present, and partly on the cost and resource implications.
- **Efficacy of detection:** It should not be assumed that, having selected a plant or plant product to inspect, an inspector will always detect a disease if it is present. This can be for several reasons. For example, symptoms may be difficult to spot visually because the diseased area is high in a tree canopy, the light levels may be poor, conditions may be windy, or there may be no symptoms to see due to latent infection. Alternatively, failure to detect a disease may simply be the result of human error. It is possible to measure these difficulties, which can help determine the sample size needed to minimize the risk of error at an affordable cost.
- **Sample size:** The number of plants to inspect or the number of samples to be taken—the sample size—is determined by decisions taken about confidence level, detection level, the efficacy of detection, and the tolerance level before action is taken. Tables exist that provide the mathematical basis for sample size determination and NPPOs consult these when planning a specific survey.
- **Sample unit:** Before decisions can be made on sample selection, we have to decide what a survey is going to inspect. This could be anything from a wood of huge and ancient trees, or five hectares of potatoes, to individual leaves or plants. This decision is affected by how evenly and easily a disease is spread, and what parts of the plant it affects.
- **Tolerances:** A decision to take action—for example start eradication action, reject a consignment, or downgrade a certified crop—will be based on risk assessment. How much tolerance to allow—if any—of diseased material in a crop, woodland, or nursery has to be decided before the sampling process begins. NPPO often set a zero tolerance for regulated diseases, because of the damage they can cause if they get established in an area or country.

How a survey is carried out is largely decided by the sampling method; it also depends on the level of accuracy required. Does it need to tell us merely that a disease is present or absent, or—if present—does it need to give us some idea about population size and frequency?

If a survey needs to be statistically supported, the common methods are simple random, systematic, stratified, or cluster sampling.

In *simple random sampling*, each sample unit must have an equal probability of being selected for inspection. This is often achieved by using random number tables, for example the table would instruct an inspector to select the 5th, 11th, 25th, 28th, 37th . . . plant on a growing bench (Figure 6.8). In practice this method is very difficult to achieve, because it needs more people and therefore costs more money, and it can be very difficult to access the plants in this way. It is also not the best method to use when disease is not randomly distributed through the plants.

In *systematic sampling*, on the other hand, sample units are selected at fixed intervals. For example, every twentieth tree in an orchard is inspected.

Stratified sampling involves dividing what you are looking at into separate subdivisions or strata and then selecting sample units from each and every subdivision. The selection can be either by simple random or systematic sampling. An example would be to select a box of apples from each layer on a pallet (strata) and then inspect every tenth fruit in those boxes.

Finally, in *cluster sampling* samples are taken from certain areas only or at certain times of the year, for example inspecting Japanese Larch plantations in Cornwall, UK.

When a survey is simply for detection we can use methods that can provide exactly the information required—but no statistical interpretation can be made from the findings regarding disease levels. A number of alternative non-random methods are available.

We can choose *convenience sampling*, when the surveyor selects the most convenient sample unit without using a simple random or systematic method. This convenience can be due to accessibility (or otherwise) of the plants, speed, or cost.

Figure 6.8 Which plant do you choose to inspect?

Pradana/Shutterstock.com

Another alternative is *haphazard sampling* whereby the surveyor arbitrarily selects sample units. However, while the selection may appear random to the surveyor, it can easily be subject to unconscious bias, and so will not be truly random.

Finally we can use *targeted sampling*. This method is common when the surveyor is trained in crop production and the disease's biology. The inspector will choose the areas to inspect in a way which maximizes the chance of finding a quarantine disease. So, for example, when inspecting ornamentals grown in glasshouses or polytunnels, inspectors will target sections orientated towards the south (warmest area), damp areas near water leaks, areas near heating pipes, areas close to vents, areas without air movement, and areas around broken glass or torn plastic. Their expertise greatly increases the chances of identifying a potentially damaging disease. Because the sampling is targeted and statistically biased, you can't use these findings to make statements about the level of disease infection. It is a valid method ONLY when the purpose of the survey is detection of disease.

Case study 6.1
Consider the humble spud

Globally, potatoes have become the third most widely grown staple crop—high in carbohydrate, low in fat, and relatively easy to grow. In the UK, despite modern tastes leaning to alternatives such as rice and pasta, the potato is still a very important crop, with over 120 000 ha grown. Potatoes may be a key crop but they are also very vulnerable to a range of diseases. Given the risk to food security, much effort is put in by NPPOs to ensure potato crops remain free from quarantine disease. Surveys play an important part in this protection.

To design a good survey, you need to know the quarantine disease threats, pathways of infection, and the ways the crop is grown (husbandry), harvested, and marketed.

How do potatoes grow?

The potatoes we harvest to eat, called ware potatoes, are a highly adapted part of the stem, called a tuber, designed for food storage and vegetative reproduction. Selected tubers, called seed potatoes, are planted between February and May, with the majority going in the ground in March and April. The soil around the planted tubers is then ridged up to ensure well-drained and aerated conditions as well as raising soil temperatures to promote early growth (Figure A). The practice of ridging also ensures tubers are well shaped and evenly sized.

Figure A Immature potato crop, showing how they are grown in ridges

Bildagentur Zoonar GmbH/Shutterstock.com

Potatoes need a lot of water, particularly when tubers increase in size (called bulking), so many crops are irrigated. Some diseases are affected by the amount of water. For example, a moist soil when tubers start to form reduces common scab development, whilst too much water late in the growing season can lead to problems with powdery scab (Figure B). In addition, by avoiding soil moisture fluctuations a grower can prevent bulking issues such as poor tuber shape and growth cracks.

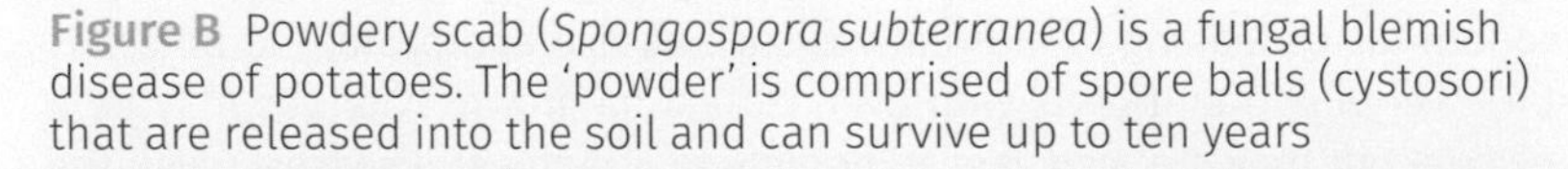

Figure B Powdery scab (*Spongospora subterranea*) is a fungal blemish disease of potatoes. The 'powder' is comprised of spore balls (cystosori) that are released into the soil and can survive up to ten years

Grandpa/Shutterstock.com

Once harvested, potato tubers can be sold direct from the field or put into store for periods lasting from several weeks to almost a year. UK farmers do not have the capacity to grow all the potatoes eaten in the country. Consequently, potato imports are also significant, with between 200 000 and 750 000 tonnes of raw potatoes coming into the country, depending on the size of the UK harvest. These potatoes can come from the EU or non-EU Mediterranean countries such as Egypt and Israel, but are prohibited from all other countries.

Potatoes grown in the UK are under threat from a range of regulated diseases which can be introduced in a wide variety of ways. Routes of introduction include the irrigation water, the soil, insect vectors, and in imported potatoes, both those imported as seed potatoes, which are planted to grow as a crop, and potatoes imported for use as food.

For a survey planner this presents multiple challenges in timing, location, and method.

Survey design challenge 1: Calculating the population and how many plants to inspect

To ensure the production of high yields of marketable potatoes many farmers use healthy, high-quality seed potatoes. Seed potato production in the UK is highly specialized and accounts or over 10% of the planted potato area in the country, centred mainly in Scotland and the north of England.

So important is the requirement for healthy, high-quality seed potatoes that many countries run certification schemes to officially govern their production. Under these regulations in the UK, it is an offence to market any seed potatoes that have not been officially certified in accordance with specified requirements.

To produce enough potatoes for food production, seed potatoes (starting with disease-free tissue-cultured material) need to be grown and replanted over many years to multiply the number of seed potato tubers. These are referred to as field generations.

However, each field generation exposes the crop to the potential to become infected with disease.

In the UK there are three categories of seed potatoes and seven grades, which reflect the number of field generations they have undergone (Table 6.1).

The classification starts with Pre-basic, which uses small amounts of tissue cultured material to produce the first crop of tubers; Basic grades further multiply this very healthy material, before finishing as Certified grades.

Table 6.1 Seed potato classifications

Category	Grades
PRE-BASIC	Tissue culture (TC) and field grown
BASIC GRADE	S, SE, & E
CERTIFIED GRADE	A & B

As part of the scheme there is a need for an official inspection of the potato crop whilst it is growing, which provides a good example of the techniques of a monitoring survey and using tolerances. Eligibility for each grade is dependent on the number of field generations and the level of disease detected during this inspection.

Due to the fact that the inspections are part of a scheme, the 'why' and 'where' questions of survey design are already answered, as each crop entered for certification must be inspected. The timing of the growing season inspection—the 'when'—is aimed at the point when adjacent plants have met along the rows and plants are just about to meet between the rows, and at the start of flowering.

The growth stage chosen has to be late enough for it to be possible to identify the variety of potato and to pick up symptoms of disease; at the same time, it must be early enough to allow for two inspections before the crop grows so big it falls across the ridges, at which point a second inspection becomes impossible.

For the purposes of this case study we will just consider inspection for, and calculation of, virus levels in a seed potato crop.

Seed potato crop inspections are quantitative surveys, which work to an internationally agreed confidence level of 95% so the tolerance levels for viral infection for the different grades of seed potatoes become the detection levels needed to make the classifications. This in turn affects the sample size that is inspected.

With the exception of Certified grades, two inspections are undertaken and calculations are based on the detection level during a single inspection. The rationale behind adopting this approach is that the virus may not be evident at both inspections, as it is possible for the crop to be infected with virus via aphid spread between inspections.

A small amount of infection is allowed as it is almost impossible to eliminate these diseases completely. Table 6.2 shows the tolerances each grade has to allow a crop to be passed.

The inspection, by highly trained official crop inspectors, is a visual assessment of individual plants that looks for symptom expression of *Potato Leafroll virus* (PLRV), *Potato virus Y*, and *Potato virus A*, as well as other mosaic virus (Figure C).

The first step in calculating the virus level in a crop is to work out the plant population. To do this, the inspector first counts the number of potato plants

Table 6.2 Potato certifications.

	Pre-basic	Basic			Certified	
	PB	S	SE	E	A	B
PLRV, Y, and A only	0%	0.02%	0.1%	0.4%	2%	6%
All other mosaic virus	0.1%	0.2%	0.5%	0.8%	2%	6%

Figure C Look how subtle virus symptoms can be and try to imagine looking for them in a crop on a windy, sunny day

Nigel Cattlin/Alamy Stock Photo

in a 10 m length of ridge three times and takes an average. Next the inspector works out how far the ridges are spaced apart. Using these two figures the inspector can then calculate the number of potato plants in a hectare as shown in this example:

Example 1

- Three 10-m counts recorded 44, 41, and 41 plants, with an average of 42.
- A row width of 0.92 m.
- $1/0.92 \times 42 \times 1000 = 45\,652$, rounded up to 46 000 plants per ha.

As a general rule, the smaller the tolerance, the larger the sample of plants that will need to be examined to detect any virus and provide a statistically valid result.

To avoid many different separate sample sizes the required numbers are rounded up to make instructions clear, understandable, and usable in field conditions; some over-inspection is accepted. Required sample sizes for Basic grades of seed potatoes are shown in Table 6.3.

Table 6.3 Required sample size for basic grades.

	S and SE	E
Stocks up to 4ha	20 000 plants at each inspection	10 000 plants at each inspection
For each additional ha or part thereof	5000 plants	2500 plants

Figure D Inspector checking rows of potatoes

Olexandr Panchenko/Shutterstock.com

The way potatoes are grown makes it impracticable to use random tables to select plants. Instead, inspectors use a modified systematic sampling technique: they walk between two rows examining plants in both rows at the same time and repeat at appropriate intervals across the field until the correct number of plants have been inspected (Figure D).

The following examples show how an inspector works out how many rows they need to inspect.

Example 2

6.5 ha of seed potatoes to be certified at SE grade with a population of 40 000 plants/ha.

- Total plant population is 260 000 plants.
- Sample size is 35 000 plants: 20,000 (4ha) + 15 000 (2.5ha).
- 35 000/260 000 = 0.136, which is a ratio of 1:7.4.

Therefore the inspectors examine two rows with an interval of 15 rows—that is, they inspect two rows and count 13 rows before starting again.

Survey design challenge 2: Deciding on the confidence and detection level

Potato ring rot caused by the bacterium ***Clavibacter michiganensis*** ssp ***sepedonicus***, mentioned in Chapter 3, is a serious regulated disease that is notifiable in the UK. The disease could easily establish in the UK's climate, with yield losses caused by tubers rotting (Figure E) being as much as 50%—with high control costs for farmers and Government.

Figure E Close up of bacterial ooze emerging from the vascular ring of a tuber infected with ring rot

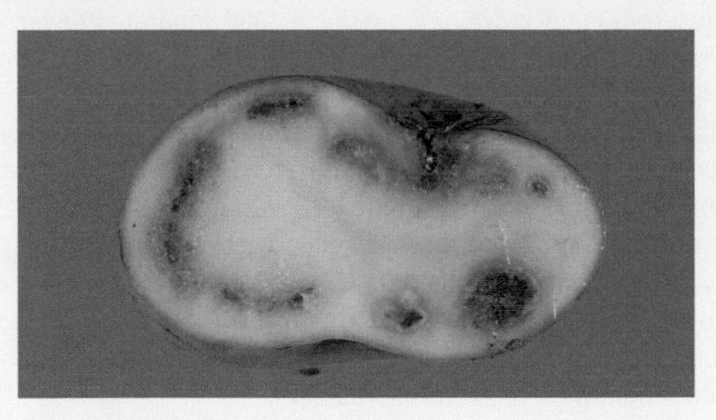

UK Crown Copyright - courtesy of Fera

If this disease became established in the UK, not only would individual crops be put under threat, but it would also prevent the UK from exporting seed potatoes.

The disease is spread mainly via infected seed potato tubers and can pass through one or more field generations without causing symptoms. Indeed, latently infected tubers are an important means of spreading the disease.

Ring rot is known to occur in a number of EU Member States, with occasional outbreaks in the UK in 2003 and 2013 at a very small number of farms having been linked to infected seed potatoes from the EU.

The disease is not easily detected in a growing crop so ring rot surveys concentrate on tuber inspections. Such surveys provide an example of balancing risk against available resources.

Table 6.4 shows the approximate number of tubers that must be tested to give a 95% probability of getting a positive result at different levels of infection.

As the survey designer is unlikely to have infinite resources, the number of tubers to be tested must be decided. Detecting to a level of 0.05% costs 18 times as much as detecting to a level of 3%. The figure is not 60 times higher as there are economies of scale—for example, travel, courier, and administration times are the same whatever the sampling rate.

In England and Wales the NPPO targets its detection survey at post-harvest inspections of seed potatoes grown, and pre-planting checks of seed potatoes from EU Member States. These potatoes are already subject to certification inspections and are deemed to be of a lower risk, so the rate of sampling is 200 tubers.

Table 6.4 The approximate number of tubers which must be tested to give a 95% probability of getting a positive result, at different levels of infection

Level of infection	3%	1.5%	0.6%	0.3%	0.15%	0.07%	0.05%
No. of tubers	100	200	500	1000	2000	4000	6000

In England and Wales during 2016 the NPPO was notified of over 2500 consignments of seed potatoes imported from EU Member States comprising many varieties from many growers. From these, 1000 separate grower/variety combinations were identified and tested. In addition, nearly 900 stocks of seed potatoes grown in England and Wales were also tested.

Another factor influencing the cost of surveying is the relatively short inspection window available: many seed potatoes arrive from EU Member States in February and March, just in time for the planting season.

During an outbreak it is important not only to identify infected crops, and the level of infection, but also to be very confident those crops deemed disease-free do not pose a risk. Given the greater certainty needed during an outbreak, 4000 tubers are tested per stock; during the 2003 single-farm outbreak, 164 000 tubers were tested from 41 stocks.

Survey design challenge 3: Deciding on survey timing, sampling frequency, and location

Potato brown rot, caused by the bacterium *Ralstonia solanacearum*, is a serious regulated disease that is notifiable in the UK. Like potato ring rot, if it became established in the UK it would not only affect individual crops but would also prevent the UK exporting seed potatoes.

Sporadic isolated outbreaks in ware potato crops have occurred in the UK over the past 25 years. All except one in 2010, which was linked to imported seed potatoes, have been associated with contaminated water and woody nightshade (*Solanum dulcamara*) as described in Chapter 3.

Current import inspections make the introduction of infected ware potatoes now very unlikely, but just to be sure the NPPO conducts an annual survey of selected watercourses in England and Wales.

If found, irrigation and spraying of potato crops is prohibited unless it uses an approved method to remove the bacterium, such as abstracting water during the winter and keeping it in storage until needed.

This targeted detection survey illustrates many factors such as timing, sampling frequency, and site selection.

Timing

The testing of water samples can detect as few as two viable pathogen cells per cm^3 of river water. However, the number of bacteria in the water is very temperature dependent, so varies according to the time of year. Once the water temperature drops below 14°C the number of bacteria falls below detectable levels. Consequently, any sampling must be in the warmer months, between July and September (Figure F).

Frequency

At a distance from infected woody nightshade plants, the bacteria can be diluted below detectable levels. This dilution effect may be higher following wet weather and associated higher flow rates. Consequently, a sample is taken every week for four weeks to ensure a cross-section of weather patterns.

Figure F Seasonal population dynamics of *R. solanacearum* in river water at a single point downstream from infected *Solanum dulcamara*

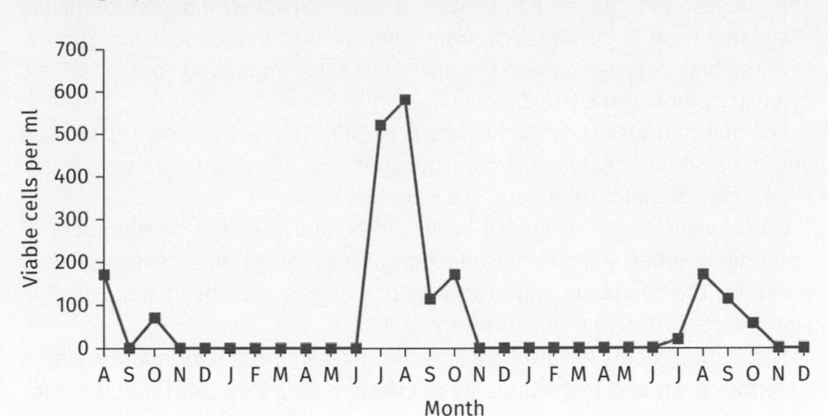

Site selection

As potatoes are grown in areas suitable for them, there is little point in sampling all watercourses in the country, so the survey is targeted to those sections that run through potato-growing areas. It is a rolling survey, looking at a number of watercourses each year, to build up a picture across the country.

The dilution effect also means that it is more likely to find the bacterium in small tributaries and ditches colonized with infected woody nightshade than in larger watercourses.

Samples are taken from two sampling points from the selected watercourse (Figure G).

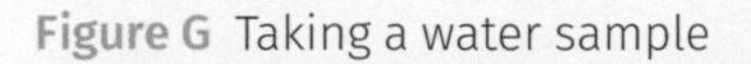

Figure G Taking a water sample

AlisLuch/Shutterstock.com

Survey design challenge 4: Deciding on sampling method

The fungus that causes Potato wart disease, ***Synchytrium endobioticum***, originates from South America, from where it was introduced into Europe, with the first recorded case in the UK in 1896. You can remind yourself of this disease by looking back to Chapter 2.

The only cultivated host is the potato. In older tubers the eyes are infected and develop into characteristic warty, cauliflower-like protuberances, which eventually rot and disintegrate (see Figure 2.9).

Due to legislation introduced in 1923, 1929, and 1973, the development of immune varieties, and the introduction of a certification scheme for seed potatoes, the economic importance of this disease has been reduced. The last reported UK field outbreaks were in 1985.

The resting spores released are able to survive in the soil for many years, so once they are found, restrictions are put in place for a very long time. These restrictions are called scheduling, and include a prohibition on growing potatoes on the land. Eventually a landowner may decide they want the restrictions lifted and a detection survey and testing are needed to accomplish this.

In the UK, the NPPO follows an international protocol whereby land can be descheduled after a minimum of 20 years since the last infection, provided that it is sampled, tested, and found to be free from viable sporangia and from any evidence of infection.

The scheduled land is divided into 0.33 ha units from which a 21-kg sample, made up of 60 subsamples, is taken. The inspector takes the samples randomly across the 0.33 ha unit using an auger to a depth of 20 cm. To achieve random sampling, inspectors walk a 'W' across the unit, as shown in Figure H.

Figure H A W-shaped sampling pattern

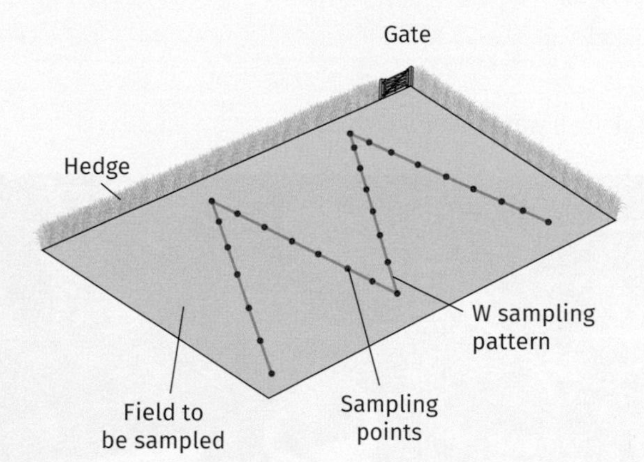

Pause for thought

Think about a garden, wood, or field near you. Can you think about a regulated disease that might affect one of the plants growing there? How would you set about designing a survey to find it? What factors would you consider?

The future

In many respects, surveying is still very traditional, in that an inspector visits and looks for visual symptoms, or takes random samples. Ensuring the best use is made of an inspector's time and expertise, as well as disease detection, is at the heart of improving survey design and implementation. Innovations include the increased use of Geographic Information Systems (GIS), remote inspection, and in-field diagnostics.

Geographic Information Systems and remote inspection

GIS software has been available for many years, and allows surveyors to analyse spatial data and find ways to present this data more effectively. GIS can help plan surveys by identifying potential sampling sites.

For example, the UK Forestry Commission (FC) uses remote inspection. Helicopters are used to put inspectors above larch trees looking for *Phytophthora ramorum*. The FC's use of GIS was integral to the planning and delivery of both inspections on the ground and the aerial survey. They had access to data of plantings from the Public Forest Estate (PFE) and added data on plantings in private woodland, which they used to identify and select sites for inspection.

Inspectors were trained in aerial survey operations, focusing on working with pilots, gaining a better understanding of the operating environment, and on practising emergency procedures and protocols.

More recent developments in remote sensing include the use of satellites and Unmanned Aerial Vehicles (UAVs)—known more commonly as drones. The technology used is always a trade-off between resolution and the area that can be covered.

The choice between satellites and drones will depend on need, but both allow for the use of specialist cameras and sensors, fir example multispectral, hyperspectral, and thermal cameras. These cameras and sensors are starting to provide accurate identification and mapping of host plants and will in the future even allow for the possibility of disease detection (Figure 6.9).

The use of UAVs (Figure 6.10) brings its own problems. They require inspectors to be trained as pilots by the Civil Aviation Authority (CAA) but they do allow inspectors to get a camera into the air, which, for example, could allow for the inspection of tree canopies or crop locations, or the identification of poor growth areas.

In-field diagnostics

As diagnostics improve, increasingly the lab is coming to the field. Besides shortening the time for a diagnosis, in-field diagnostics are also useful for the surveyor who is undertaking targeted visual inspection.

Symptoms may be similar for a range of diseases, but immediate diagnosis allows inspectors to differentiate in real time as well as allowing them to improve their symptom recognition as they 'get their eye in'. Even if the diagnosis is only down to genus level, it can act as a filter to reduce the burden on the diagnostic laboratory. For example, it avoids sending in symptoms caused by unrelated conditions, including physiological symptoms.

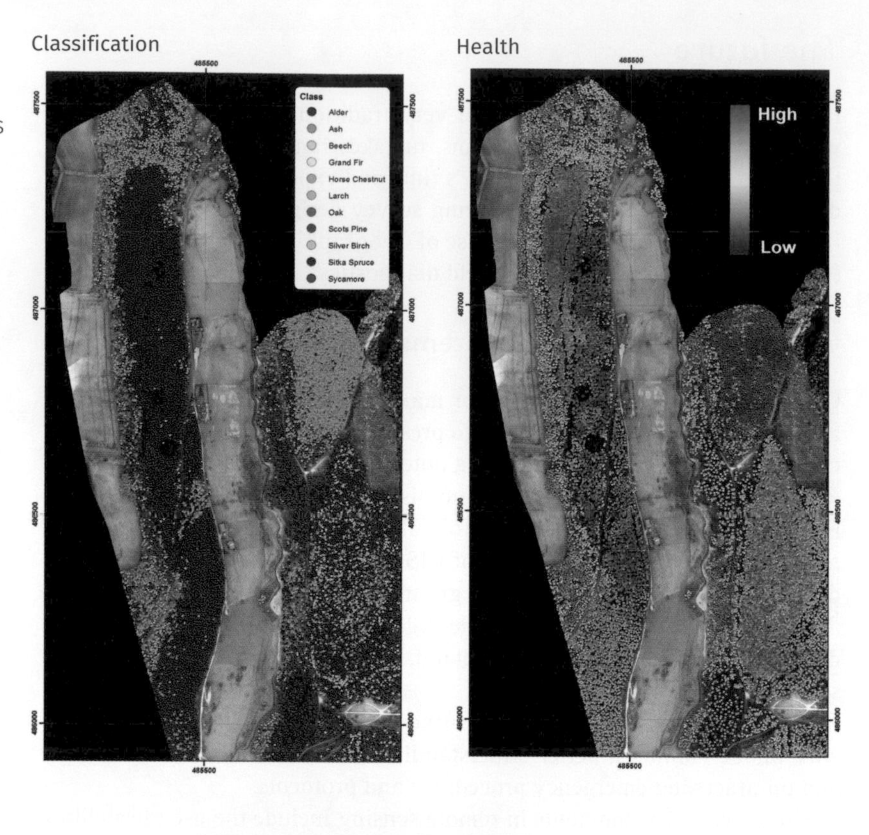

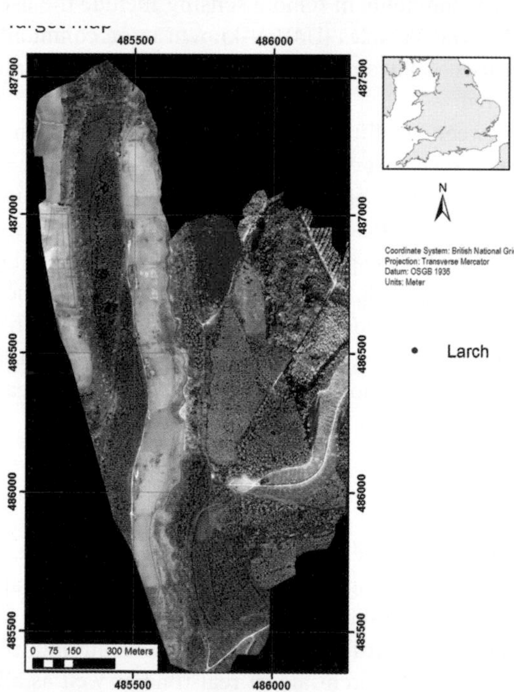

All images: FERA Science Ltd

Figure 6.9 Using aerial technology to identify disease in larch. In the panels, Classification refers to what species of tree is present, Health refers to how healthy they appear to be, and the Target map is showing inspectors where possibly unhealthy trees of the species they are concerned about are present.

Figure 6.10 A UAV—or drone—in action checking for plant disease in a crop

Andy Dean Photography/Shutterstock.com

Two examples of in-field diagnostic technologies are Lateral Flow Devices (LFD) (mentioned in Chapter 2) and loop-mediated isothermal amplification (LAMP) machines. LFDs detect the presence or absence of a target analyte (e.g. a pathogen) using immunoassay methods; the most common example of this technology is the home pregnancy test. LAMP machines exploit a relatively new DNA amplification technique that, due to its simplicity, ruggedness, and low cost, makes field applications possible.

If an inspector can identify the presence of a pathogen in the field, appropriate control measures can be put in place, if needed, extremely quickly. These new technologies offer the exciting prospect of faster, more effective surveillance in future. This is turn can lead to more effective control of plant diseases, enhancing global food security and environmental well-being.

Chapter summary

- Surveillance is an official process which collects and records data on pest presence or absence by survey, monitoring, or other procedures.
- Surveillance is key to preventing the spread of disease in crops, nurseries, and ecosystems.
- General surveillance information is collected from many diverse sources to alert us to the presence of new and emerging diseases.
- The objective of specific surveys is either to detect the presence or confirm the absence of a particular disease or, where present, determine how widespread it is.

- The location and timing of specific surveys needs to be decided, considering factors like geography, climate, and the biology of the host, disease, and vector where appropriate.
- The way in which a survey is carried out is determined by its objective and the level of accuracy required.
- New technologies are being used to help surveyors to better target their inspections and improve disease detection.

Further reading

https://www.ippc.int/en/publications/615/

International Plant Protection Convention, International Standard for Phytosanitary Measures, ISPM 06 Guidelines for Surveillance, 14 January 2016.

https://assets.publishing.service.gov.uk/government/uploads/system/uploads/attachment_data/file/307355/pb14168-plant-health-strategy.pdf

Defra's Protecting Plant Health report.

http://www.plantsentinel.org/index/

An example of a source of general surveillance is the International Plant Sentinel Network.

https://www.opalexplorenature.org/surveys

Find out about citizen science projects run by OPAL.

https://potatoes.ahdb.org.uk/

The Agriculture and Horticulture Development Board provides information on potatoes.

http://chalaramap.fera.defra.gov.uk/

Investigate any UK location to see if Chalara dieback of ash has arrived in your area.

http://www.agriculture.gov.au/pests-diseases-weeds/plant

Check out Australia's hotline and top 40 pest and diseases.

Discussion questions

6.1 What information does the general public need and how would you deliver it to encourage them to report new diseases?

6.2 Describe an outbreak scenario for the disease that is of most concern in your country and summarize the factors you need to consider in designing a survey.

6.3 Summarize the methods that surveyors can use to find new diseases.

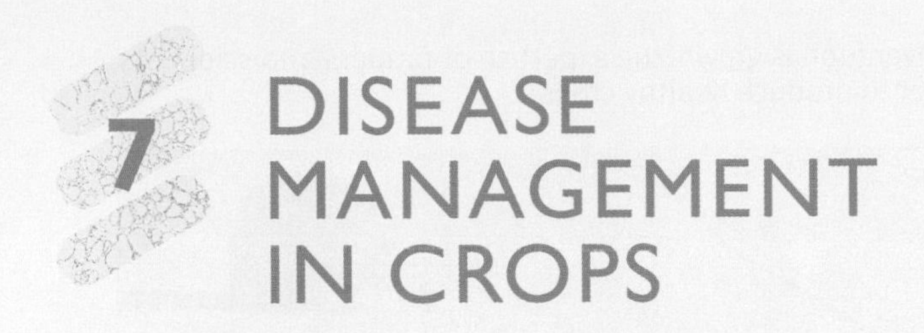

7 DISEASE MANAGEMENT IN CROPS

Dr Julian Little

The previous chapters have illustrated how fungi, viruses, and bacteria can all infect plants and result in them getting 'sick'. You have met some of the problems these diseases cause for farmers who need to get the best yield and best quality from their crops to be profitable, for the billions of people who rely on plants for everything from food and clothing to building materials and energy, and for ecosystems around the world—see Figure 7.1.

Growers have many tools in their toolbox to help them reduce the impact of pathogens on the crop plants we all depend on. You might think that the first line of attack against plant diseases is the application of chemicals such as fungicides to kill the pathogens or insecticides to prevent insect vectors from transferring diseases from plant to plant. However, in the same way that medicinal drugs are not the first course of action when we are ill, there are many other considerations *before* chemicals are applied to crops. Many of these alternative techniques rely on a grower's experience and knowledge of their farm, and on scientific knowledge of the pathogen itself.

Chapters 5 and 6 explained the importance of biosecurity and surveillance in protecting plants against diseases. Once pathogens get into the trees that make up our forests and woodlands there is little we can do, apart from felling, burning, and surveillance. But when it comes to crop plants, the stakes are in some ways even higher: if the food crops fail, people may die of starvation. In this chapter we will focus on the ways we manage and treat diseases that affect the crops we all rely on.

Farmers use a whole range of techniques carefully to ensure their crops are healthy, to reduce the risk of infections occurring, and to treat diseases or pests when they do occur. This holistic approach is known as Integrated Crop Management (ICM); its key components are explained further in Chapter 8.

Figure 7.1 Wherever food is grown, the expertise of farmers and scientists can work together to produce healthy crops

Juice Images/Alamy Stock Photo

AGL Photography/Shutterstock.com

Cultural control

It is easy to underestimate the impact different growing systems can have on disease control. Measures including planting seeds later or earlier, ploughing, crop rotations, and break crops can have an enormous effect on the health, and therefore the yield and quality of crop plants. These measures are all forms of cultural controls—the management and reduction of plant diseases based on the way the plants are cultured (grown).

The most basic method of cultural control is crop rotation—growing a series of very different crops from different plant families in a particular field over a series of years. The idea of letting land rest or lie fallow every few years is thousands of years old. This break in cultivation both helps to maintain field fertility and reduce plant disease, but also has less obvious benefits in areas such as weed management. Four-field rotation, when a sequence of four crops (typically wheat, turnips, barley, and clover) were grown in sequence in four fields, was developed in the sixteenth century in Belgium and made more widely popular by the British agriculturist Charles 'Turnip' Townshend, seen in Figure 7.2, in the eighteenth century. Cultural control has come a long way since then, but crop rotation is still an important feature.

There are a number of key points to highlight when considering cultural ways to control plant diseases and reduce crop yield and quality losses to pathogens.

Figure 7.2 The 2nd Viscount Townshend, also known as Charles 'Turnip' Townshend, may not look much like a farmer—but his ideas on crop rotation revolutionized agriculture

World History Archive/Alamy Stock Photo

Know your enemy

Knowing how a pathogen survives and spreads within and between crops gives clues as to how infections can be reduced or stopped. These all feed into cultural control measures.

Where does the disease come from?

The pathogens that cause plant diseases arrive in a variety of ways, including infected seed, insect vectors, air currents from distant crops, water droplets from weeds, volunteers (plants which grow from seed left in the ground from the previous crop), or crop debris. Farmers can't stop the wind blowing from other fields or even from other parts of the world, carrying spores in the air currents, but they can try to reduce airborne spread by careful planting. If the spores are originating from weeds, crop debris, or volunteers, then reducing these can reduce the spread of the pathogens when a new crop is sown, and also increase the length of time before the disease is seen in the crop.

What are the hosts of the pathogen?

Some pathogens can live on many different plants, whereas others are very specific, infecting only one species. Knowing the host range of the pathogen can be a useful way to stop the cycle of reinfection. Growing different crops in a rotation is the first step to breaking the disease cycle by reducing the build-up of infectious material in the field.

An example of where crop choice in the rotation is important is a disease caused by the fungal pathogen, *Fusarium*. This fungus can affect the base of the stems of the wheat crop and then be transferred to the grains in the ear where it causes Fusarium Ear Blight (FEB), which you can see in Figure 7.3. In wheat this causes the grain to shrivel, reducing the yield. But there is an added problem: **mycotoxins** are produced by the fungus in the grain. These toxins are harmful to animals, including people, and grain containing the toxins cannot be sold for human or animal food.

Figure 7.3 You can see the difference between the healthy and the blighted ear of wheat. Anything that reduces the size of the individual grains reduces the yield of the crop.

tetiana_u/Shutterstock.com

Nigel Cattlin/Alamy Stock Photo

FEB is not caused by a single fungal species but by a group of related species. *Fusarium culmorum* and *F. graminearum* produce the most serious mycotoxins and are heavily regulated. In recent years, the production of maize for anaerobic digesters has increased and, with this, there has been an increase in the incidence of *F. graminearum* in wheat, where maize is also grown in the rotation. The same pathogen is colonizing the maize but because the maize is used for fuel, not food, it is less of a problem in this crop. The pathogen stays in the maize stubble (the short bits of stem left in the ground after harvesting) and crop residues, providing inoculum for the next crop of wheat, where it can ruin the crop. This illustrates the importance of knowing the host range of the pathogen to ensure changes in the rotation do not result in more of a problem.

Two other factors that are closely linked to cultural control include choosing varieties which are resistant (or partially resistant) to diseases, and treatments using chemical or biological control products. You will find out more about these in later sections of this chapter.

Case study 7.1
Septoria wheat blotch

Septoria is a damaging fungal pathogen of wheat and can result in significant yield loss of up to 50% if the weather conditions are favourable for it. What's more, the quality is also affected as the grains from infected plants are often smaller. Most of the wheat crops in the UK will have some level of disease on the leaves. The flag or top leaf provides most of the energy to fill the ears of wheat. A Septoria disease **lesion** that covers just 1% of the total leaf area of this leaf results in a 1% reduction in yield in the grain harvested. You can see Septoria lesions in Figure A. When this is multiplied across a field, you can easily see how large reductions in yield can occur. So, how can we stop Septoria?

Firstly by knowing the enemy: what conditions does the disease need to thrive? Septoria has been the subject of research by universities, research institutes, and agrochemical companies around the world for many years, as it is such a devastating disease. By using this information we can make small adjustments to how we grow the crop, which makes it harder for the fungus to infect the plants. Once a grower has established that their fields are in a high-risk area for Septoria, they can look for ways to reduce that risk.

Variety resistance or tolerance is a key factor, as are chemical fungicides (described in detail in later sections of this chapter), but cultural control has a significant impact. Septoria is an excellent example of how knowledge about the pathogen can be used to change the way the crop is cultured, in a bid to reduce the impact of the disease in a real-life setting. These include the following.

Figure A If you look at any wheat crop in the UK it is highly likely that you will find some lesions on the lower leaves. Lesions may be present on the higher leaves too, depending on how well the grower has controlled the disease.

Vilor/Shutterstock.com

Being aware of conditions likely to increase the risk of disease

Septoria is referred to as a wet-weather disease: moisture is needed for the spores to be spread between plants and from one leaf to another. Septoria is a problem in most temperate maritime climatic zones such as the UK. However, disease levels depend on the weather conditions: the damper the conditions, the worse Septoria will be. The grower uses his knowledge of the climate on his farm to identify whether Septoria is going to be a big risk to his crop, or a smaller one. The grower cannot influence the weather but does know whether the area in which he is located is relatively wet (e.g. Herefordshire in the west) or dry (e.g. East Anglia in the East). So the losses to Septoria are likely to be higher in the west, and growers there must be especially vigilant for the disease, or may choose to grow other crops.

Of course, other diseases have different optimum conditions, so although growers in the East have fewer worries about Septoria, they are more likely to be concerned by rusts which prefer the drier conditions, and must consider the cropping system accordingly.

Changing sowing time to make it harder for the disease to infect

Septoria, like all fungi, has an optimum temperature range. For Septoria this is 15–20°C; within this range, the time between a spore infecting a plant and the next generation of spores being produced is only about 21 days. At temperatures outside of this range the cycle takes longer. By delaying the drilling of the wheat until later in the autumn, when temperatures are generally a bit cooler than with an earlier drilling, the potential for initial infection is not reduced, but the number of disease cycles the crop can support *is* reduced. This decreases the amount of inoculum available, so slowing disease spread.

In addition, Septoria infects the plants either through the stomata or by directly penetrating the leaf. Therefore, early drilled crops are easier to infect because they tend to be more lush than later crops, making it easier for the fungus to infect the soft leaf tissue. So, by altering the drilling date, like the farmer in Figure B, growers can influence how easy it is for Septoria to infect their crop.

Reducing visible straw as this is a major source of inoculum for the next crop

One of the best cultural practices is to reduce the inoculum level in the infected crop residue during the previous season—in effect maintaining a continuous cycle of disease control measures. A second, very effective practice is to reduce the volume of infected debris before drilling—for example, by ploughing the stubble and straw residues to bury them under the soil.

Figure B Simple measures like delaying planting the wheat seed can reduce the risk of infection by Septoria

Fotokrostic/Shutterstock.com

Stopping the green bridge as a source of inoculum for the next crop

Volunteer plants which originate from spilt grain from the previous harvest provide an excellent host for the disease during the period between harvest and the emergence of the next crop. This is referred to as a green bridge as the living material supports the growth and sporulation of the pathogen before it infects the next crop. Controlling these volunteers either by reducing the grain dropped at harvest or chemically by using herbicides before the crop is drilled can limit this transfer route for spores. One of the best cultural practices is to reduce this debris from the same plant species.

Increasing the gap between growing susceptible hosts to reduce inoculum levels

Crop rotation—growing a different crop after wheat has been harvested—can break the cycle of infection. However, in the UK it is common to grow several wheat crops after each other before a break crop is included in the rotation, because wheat is in demand and in many cases is the most profitable crop. If the break crop is a dicotyledon (such as oilseed rape) or a different cereal which is not a host to Septoria (such as barley), then there is a break in the inoculum build-up in the field.

A break year (i.e. when the host is not grown) is a valuable technique for controlling Septoria, since the pathogen inoculum in the straw debris decreases as it decomposes.

Any one of the measures listed here would result in a small change in the levels of Septoria in the crop, but when combined with growing the most resistant varieties, and the judicious application of **fungicides**, growers stand the best chance of beating the disease.

Of course, it's worth noting that this case study has only considered one disease of wheat: there are many other diseases which must also be considered. Each disease has a different host range, preferred climatic conditions, method of infection, and so on. As such, growers must make a balanced choice of the techniques they will employ from their ICM toolbox to best suit their individual circumstances.

? Pause for thought

Consider the conditions and cultural controls that could result in both the worst-case and best-case scenario for Septoria infection in a wheat crop.

Discuss the differences in the problems of dealing with fungal diseases which will face farmers growing wheat in the UK compared with farmers in less economically developed nations.

Chemical control

To achieve-good quality yield and therefore income for the grower, most crops will receive chemical applications to control disease. These have been used for centuries to help ensure there is a good crop to harvest. In 1000 BCE,

Homer (a Greek poet) wrote about the use of sulfur to keep pests away from plants! In the eighteenth century, mercurous chloride was used as a fungicide and seed treatment, while nicotine was used to control aphids. As you saw in Chapter 2, by the late nineteenth century Bordeaux mixture—copper sulfate, lime, and water—was used to control downy mildew on grapevines. By the twentieth century, chemical pest and weed control was well established. Figure 7.4 summarizes some of the big steps.

The early trailblazers of chemical controls did not always understand the side effects of their treatments: using arsenic-containing compounds on crops wasn't always good for the people who ate them, for example! Today, the chemicals that growers can apply to a crop, and the timing of any chemical applications, are highly regulated. This is to maintain the safety of the crop we eat, as well as limiting any damage to the environment. Even pesticides applied to organic crops (such as copper sulphate) have to be regulated.

In the twenty-first century, chemicals to control crop diseases and pests are generally used with restraint and as part of a package of control measures (see Chapter 8). Many different types of chemicals are available to the grower. Some will be specific to particular pathogens, while others have a much wider activity. An example of the chemicals that affect specific pathogens is the **mildewicide**. These chemicals only control the mildews and have little effect on other pathogens.

The chemicals used are all **pesticides**, whether they affect fungal pathogens (fungicides), insect pests and vectors of disease (**insecticides**) or the weeds which grow and compete with a crop—and may spread disease (**herbicides**).

Figure 7.4 5000 years of crop protection

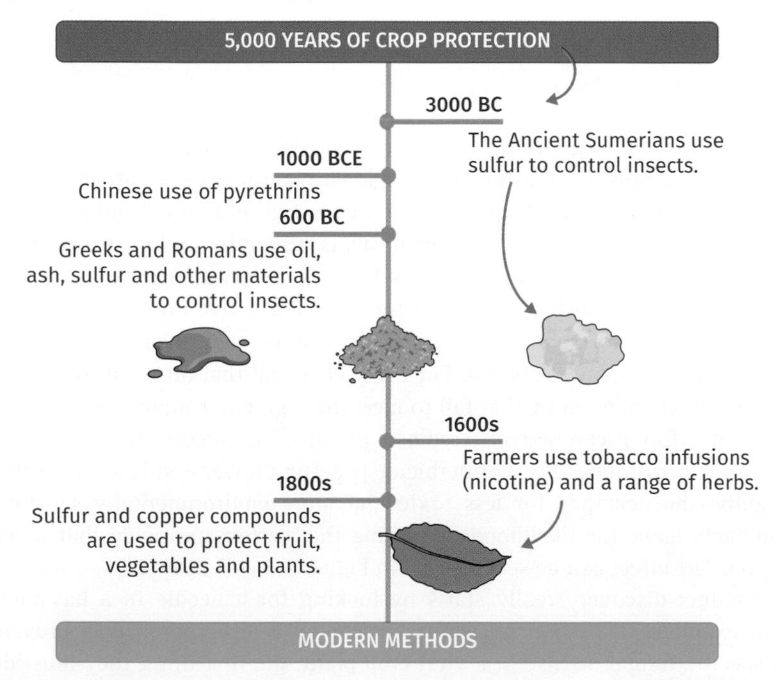

CropLife International

Crop control chemicals can be grouped by their chemical properties but are more often classified by their mode of actions—that is, the chemical process the chemical disrupts. Plant viruses are as difficult to attack chemically as the viruses that affect people, so there are no chemical treatments against viral diseases; the only way of preventing them is by the control of insects, such as aphids, which spread the viruses as they move and feed from plant to plant. Where the viral disease has already taken hold, some farmers resort to destroying the crop to try and halt the spread of the virus.

Antibiotics are not widely used in the treatment of plant diseases either. In fact, bacterial plant diseases are also almost impossible to treat chemically at the moment. This means that the great majority of pesticides used in the control of plant diseases affect fungal pathogens. A few examples of chemical groups used to protect cereal crops against fungal pathogens are given in Table 7.1; this is not an exhaustive list.

Table 7.1 Modes of fungicide action

Chemical class	Effect on fungal pathogen
Succinate dehydrogenase inhibitors (SDHI)	Work by targeting succinate dehydrogenase in fungal mitochondria, disrupting cellular respiration, and so destroying the fungus
Demethylation inhibitors (DMI)	These are made of up different chemical groups (triazoles, imidiazoles, pyridines), but all inhibit demethylation in sterol biosynthesis
Multisite inhibitors (MI)	These compounds inhibit many biochemical pathways in the pathogen

Making new chemical controls

Let's start at the beginning by looking at the development of a new chemical product. Most chemical controls against plant pathogens are developed in the more economically developed areas such as Europe and the USA. Today, the registration of new chemicals is a lengthy and costly activity; see Figure 7.6. It can take 12 years from discovery to marketing a new product, and can cost £250 million for each new active ingredient. It takes around the same amount of time as developing new medicines to treat human diseases, but costs less. For every chemical that makes it to market, there will be many more that fail to meet the high environmental standards needed before it can be registered as a product. The success rate of finding a new successful pesticide or herbicide is getting lower and lower as—quite rightly—the demands for less toxic and more environmentally sensitive products make the likelihood of finding the 'perfect' pesticide that much lower. The effect can be seen clearly in Figure 7.5 and Table 7.2.

Product discovery really starts by looking for a needle in a haystack. For example, if a company wants to develop a new chemical to prevent a specific fungal disease in a vital crop plant, the first thing they must do

Table 7.2 Number of products processed leading to a successful product launch

		1995	2000	2005–08	2010–14
Research	Synthesis	52 500	139 429	140 000	159 574
Development		4	2	1.3	1.5
Registration		1	1	1	1

Phillips McDougall, 2016

Figure 7.5 These data reflect the increasing numbers of chemicals which are screened to find one successful product

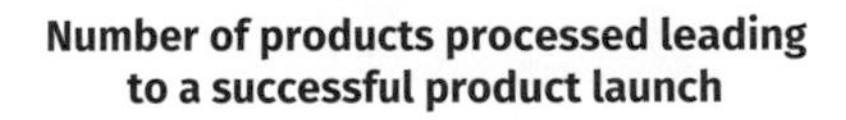

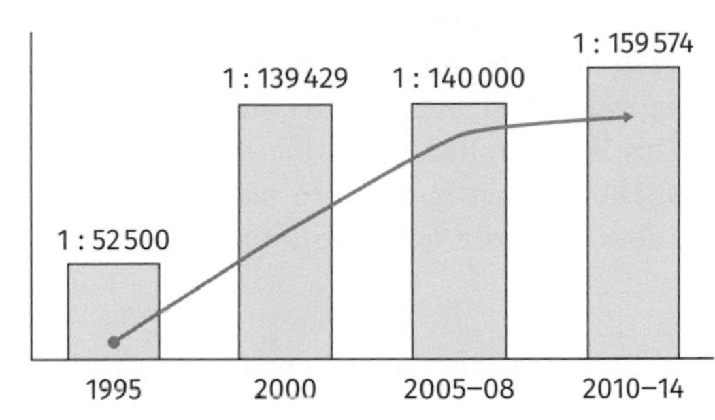

PhillipsMcDougall, 2016

is find a chemical which destroys the fungus. Sometimes this involves the chemical synthesis of a molecule that is similar to an existing anti-fungal product. Alternatively, naturally occurring products will be tested for their anti-fungal properties. If this is successful, then the active ingredient (the molecule with the fungicide activity) will be synthesized to make a pure solution.

At this stage the molecule will be studied to find out if it is likely to cause undue environmental or crop damage. If this is the case, the work is then stopped and the process returns to square one: it is back to looking for a new molecule. If the molecule has passed these initial safety tests, however, then work on the formulation begins (Figure 7.6). The active ingredient is mixed with carrier molecules. These molecules help the active ingredient to dissolve in water, keep it stable before use, and can also help the active ingredient 'stick' to the plant leaf or to penetrate the plant surface. At the same time, the carrier molecules mustn't affect the pH of the formulation and must also keep the active ingredient in a format which prevents it from losing its anti-fungal activity. They must all also be safe for the environment, for pollinating insects such as bees, and for the people or domestic animals who eat the final crop. The correct choice of formulation is key to a successful product, as poor formulation can result in the product running off the leaves or failing to reach the fungus.

Figure 7.6 Chemical controls don't just contain the active chemical—that is just part of a complex formulation to ensure it is as effective as possible

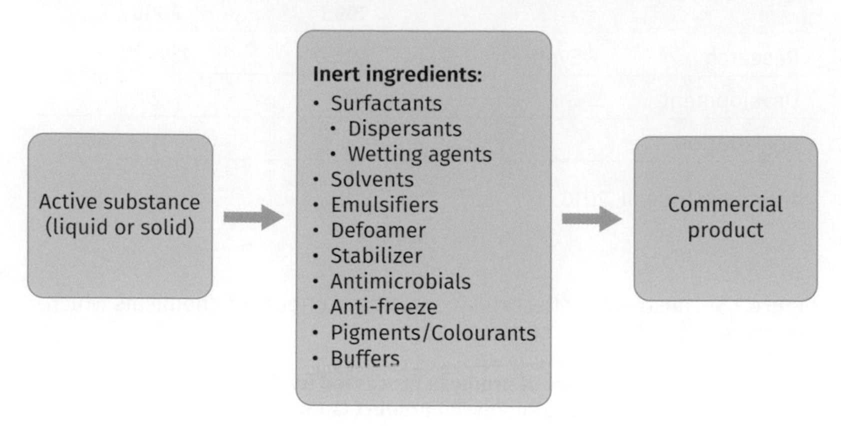

The formulation will now become the product but will only become a marketable fungicide if it passes all the regulatory requirements (see Figure 7.7). This is controlled under the European Registration Regulation 1107/2009. The product is tested to make sure it is safe to use in the environment, does not pose a risk to animals or water courses, and

Figure 7.7 It takes a long time, and input from a lot of different scientists, to develop a new pesticide or herbicide

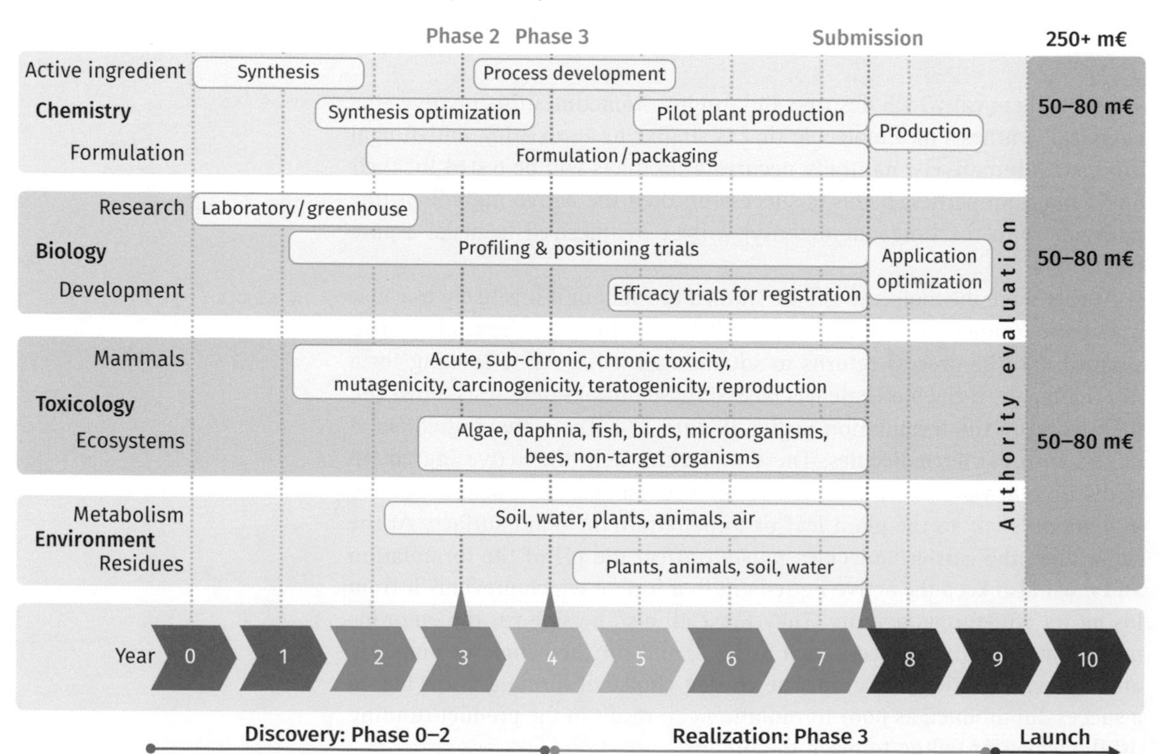

will not taint our food. All of this data is put together in a dossier which is submitted to the government for approval for the uses requested. If a product is registered for wheat it cannot be used on apples until a dossier has been submitted for that crop. Currently the European Union will assess the active ingredient and then the government of the specific country will assess the dossier and decide if the product can be registered for use in that country. This approval can take as little as one year but can take up to 2–3 years. Once a product has government approval it can be sold.

To spray or not to spray?

Around the world, farmers and growers have to make decisions about whether to apply chemicals to their crops to protect them against diseases and pests, which can ruin the crop, and risk people going hungry and lead to potential ruin for the farmers. However, these chemicals can be expensive, and they can only be applied at certain times in the growing cycle to avoid contaminating the final crops. It isn't an easy decision: a number of factors have to be considered.

Keeping it safe

Over time, people have become increasingly aware of the potential hazard of pesticide or herbicide residues building up in food chains. The classic case of the discovery in the 1970s of the impact of dichlorodiphenyl-trichloroethane, commonly known as DDT, on animals further along the food chain has made the governments who regulate the use of agricultural chemicals, growers who use the chemicals, and the scientists who develop new chemicals to control plant diseases and pests, very aware of testing the environmental safety of compounds. Figure 7.8 shows you the DDT chain.

Residues in our food aren't the only concern: any chemical used also has to be safe and not cause any problems for the processing of the harvested crop, such as making bread from wheat or beer from barley. When a product is being registered as a fungicide on barley, tests have to be done to ensure any residues on the grain will not affect the brewing process or have an effect on the taste of the beer. Only products that have been identified as not having an effect can be used on the barley crop which is destined for brewing.

Cost vs benefit

As with all things, the cost of fungicides varies depending on the active ingredients included. Growers think very carefully about how much they want to spend on the crop and also what the return (price of the yield) will be. Also, some crops can still reach a high price with some blemishes/disease present, while other crops (fruit, for instance) have to be perfect. All of these factors will influence the type of fungicides used. All fungicides are expensive, but the cost of them needs to be balanced by the amount of money growers get back at harvest when they sell the crop.

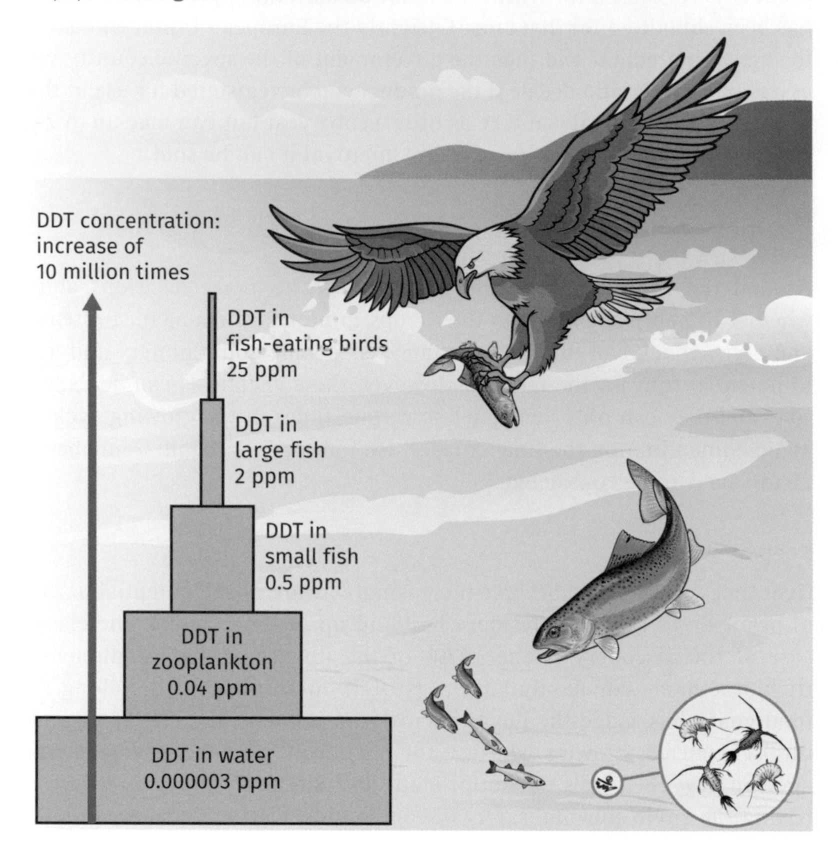

Figure 7.8 The damage done by the passage of DDT through food chains around the world has informed the work of scientists looking for new crop-protecting chemicals ever since

When to apply

Farmers want to apply fungicides at the most critical times for the crop, but not spend so much that they won't get back the costs when they sell the produce. Fungicides are also seen as an insurance policy. In the UK, some plant disease will always be present but some years will be worse than others, often depending on the weather conditions that follow.

Most fungicides work by stopping the initial growth of the pathogen. They therefore have to be applied at the first signs of the disease. It is at this point that growers will be weighing up the information available to them to make their best guess of how high the risk of a disease epidemic is. The final decision over whether or not to use a fungicide, and how much to use, will be made locally by each individual. Whether or not the right decision is reached is part science, part good husbandry, part weather forecasting—and part luck!

Which pesticide?

Pesticide choice is not a free choice. Growers will choose fungicides or other chemical controls based on what the end use of the crop will be, the cost, disease pressure in the area, and the disease risk of the crop (which will depend at least in part on the success of any cultural controls used). There are also legal restrictions to follow, and the end user may also have guidelines that determine what can and cannot be applied to the crop. The legislation surrounding the application of all pesticides, including fungicides, is very complex and is an area that is continually evolving. As a result, only qualified people can advise growers on what can be applied to the crop, in the same way as doctors are the only ones who can prescribe certain medicines.

Those who advise on the application of fungicides and other pesticides need to be BASIS qualified. This means they will have undergone initial training and will have maintained an up-to-date knowledge of the legislation and guidelines by attending workshops and meetings.

The process of what to use on the farm begins by reviewing all of the cultural controls we have already covered, the varieties grown, weather conditions, online prediction tools, and the disease risk. The grower will also think about the end use of the crop and what products he can use to be able to sell into that market. Taking all of this into consideration, a choice of products is then decided upon.

The product labels will then be studied to find out if there are any additional restrictions based on the growth stage at which a product can be applied to the crop. Following all this information the grower will devise a spray programme for his crop.

The time of day that spraying takes place can also be very important. For example, when a grower wants to spray a fungicide at the same time as an insecticide to control damaging pests, they will be keen to avoid harming bees and other beneficial insects, and are likely to spray crops early in the morning or late in the evening when pollinating insects are less likely to be active, as seen in Figure 7.9.

Figure 7.9 Pesticides are a powerful tool, and all growers need to use the chemicals available carefully to maximize their longevity on the market and minimize any damage to the environment

oticki/Shutterstock.com

Fungi fight back

Currently we are hearing a lot about antibiotic-resistant bacteria, which are a concern for doctors. In a similar way, the fungal plant pathogens largely targeted by chemical control can develop resistance or tolerance to fungicides. When resistance arises, alternative modes of action are often the only way forward. However, the discovery of new modes of actions, and the cost and time it takes to get a product to market, is considerable. Consequently, growers, manufacturers, and researchers are keen to take care of and manage the products carefully to maintain their effectiveness.

In Europe a cross-national group of specialists have come together in the Fungicide Resistance Action Committee (FRAC). This specialist technical group aims to provide fungicide resistance management guidelines to prolong the effectiveness of fungicides and to limit crop losses should the fungus become resistant to that mode of action. Members of FRAC identify which pathogens are likely to evolve to become resistant based on lab studies, pathogen life cycles, and past experience of other modes of actions. They also identify which modes of action are most at risk of resistance developing. This is based on which metabolic pathway is targeted and how complex the mutations would need to be in the pathogen in order for them to become resistant.

The FRAC group is made up of plant pathologists who are studying the pathogens in research organizations and in industrial companies. The data is shared within the group to help steward products. In addition, some countries have their own Fungicide Resistance Action Group or FRAG. In the UK, this is FRAG-UK, made up of representatives from the agrochemical companies, research organizations, and government. FRAC provides guidance for FRAG-UK, which in turn focuses on the diseases specific for the UK and the way chemicals are used. All of the data available is discussed and stewardship guidelines are developed to help maintain the efficacy of products. These guidelines are then used by government when a new product is registered, and also help to determine the label requirements, such as those in Figure 7.10.

For example, Septoria, which you met in Case study 7.1, is a disease which has many cycles in a season (from plant infection to spore release) so it multiplies rapidly if not controlled. It also undergoes both sexual and asexual reproduction. Past experience shows this pathogen has the ability to develop fungicide tolerance to a specific mode of action relatively quickly. This evidence was coupled with the release of a new class of fungicides called the SDHI, which work on a single biochemical pathway in the pathogen (see Table 7.1). FRAC pulled together this information and issued guidance, accepted by FRAG-UK, which has resulted in a statement on the label of all SDHI fungicides used to control this pathogen. This states they must only be used twice in a season and must be mixed with an effective mix partner with a different mode of action. The reasoning behind this is that a naturally occurring mutant which has less sensitivity to SDHI would be controlled by the mix partner. This is a very fine balance, as mutations are happening all the time, and sensitivity to other modes of actions is also appearing. However, so far this stewardship is prolonging the life of the SDHIs: they remain an effective defence against Septoria.

Figure 7.10 You have to read the labels carefully when you plan to use a fungicide on crops. Using them carefully will help to ensure we have chemicals to protect our food long into the future, when they are needed.

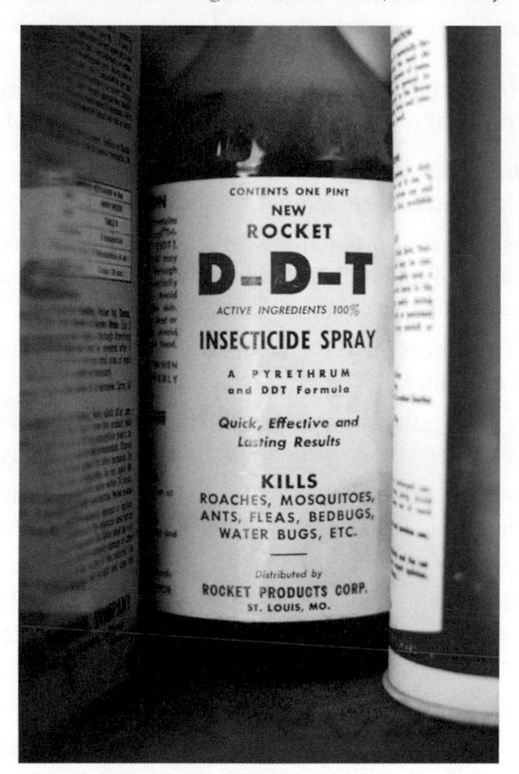

William H. Mullins/Science Photo Library

In contrast, a group of fungicides called strobilurins were not stewarded in this way and their usefulness was very quickly diminished as resistance in key fungal pathogens emerged.

Selective breeding and genetically modified organisms (GMOs)

For thousands of years, selective plant breeding was as simple as choosing the seed from the best-looking or most productive plants from one year's crop and planting them for the following year's crop. Such selection wasn't just for the plants that gave more food: in bad years, when disease levels were high, a farmer would choose seeds from those plants that were more disease-resistant than their counterparts.

Traditional plant breeding involves crossing plants that have *natural variation* (caused by spontaneous mutations in their DNA) to find new useful crop varieties. Since the 1950s, plant breeders have deliberately introduced significant variations in crop plants (*forced mutations*), initially by putting seed in a nuclear reactor, and more recently by using chemical mutagens such as ethyl methanesulfonate (EMS). In both cases, the plant breeder would grow up the plants from the resultant seeds and screen the plants for those with useful new attributes to feed into conventional breeding programmes. The disadvantage of both methods is that the mutations are random and large numbers of 'treated seeds' need to be screened to find the desired attribute in a plant.

Today's selective plant breeding still relates to that history, but we now have a huge range of tools to help us increase our chances of success:

- Gene mapping enables us to understand why the genetic make-up of some plants make them more resistant to disease than others.
- By causing artificial mutations we can produce plants harbouring a wider range of mutations in order to search for resistance genes (so-called mutation breeding).
- We can use gene markers to predict which genes are involved in disease resistance.
- With marker-assisted breeding we can direct plant breeding to encourage desired resistance genes to be present and active.
- We now have the technology to transfer a disease-resistance gene from one species of plant to another (as seen in genetically modified crops).
- Using gene editing and specific tools such as CRISPR/cas9 (see Scientific approach 7.1) we can generate more effective resistant genes within a plant itself.

In the end, however, a seed breeder has to be able to demonstrate that in real fields, with real farmers facing different soil types, weather patterns, etc., the new seed will produce a good crop even when a particular plant disease is present in the crop.

Plant breeding innovations

The use of genetic modification (GM) has very successfully introduced new genes into a crop plant. For example, the introduction of an insecticidal protein from the bacterium *Bacillus thuringiensis* into cotton resulted in plants that are resistant to the otherwise very destructive cotton boll weevil. These genetically modified crops, seen in Figure 7.11, express genes for insecticidal Bt proteins in their leaves. These proteins are toxic to many insect pests. As a result, many hundreds of millions of hectares of cotton, maize, soybean, and other crops have been protected from beetle and caterpillar attacks.

Fortunately, the insecticidal proteins are not toxic to mammals or other non-target species. Unfortunately, however, the widespread use of this technique created a strong natural selection pressure on a number of pest species, some of which have started to develop resistance to the insecticide in Bt transgenic crops, reducing their effectiveness. However, when used as part of an ICM programme alongside the inclusion of additional genes coding for different proteins found in other Bt strains, this technique is still exceptionally useful for both protecting crops and reducing pesticide inputs. You can find out more about it in Chapter 8.

Figure 7.11 These GM cotton plants are safe from the ravages of the cotton boll weevil. This protects both the plants and the livelihoods of the growers and their workers.

Mike Flippo/Shutterstock.com

In many cases, a plant breeder is more interested in modifying genes that are already present within a plant than inserting completely new ones. A series of new breeding technologies have arisen to help them do exactly that. Of course, it is possible to look for 'natural' or 'forced' mutations, but these new techniques allow a change or mutation in a gene to be directed very specifically to the gene in question. For example, if you can identify a particular gene that confers resistance to a fungal pathogen, you might want to alter the DNA sequence of the gene to increase that resistance. (You might, of course, make things worse, but that would also be an interesting observation for the plant breeder.) The technique that has revolutionized the ability of scientists to tweak the genes in this way is known as CRISPR/Cas9, which we explore further in Scientific approach 7.1.

Scientific approach 7.1
CRISPR-Cas

CRISPR-Cas is an example of a gene editing innovation already being used by plant breeders. The basis of this gene editing is summarized in Figure A. With this technique, a nuclease (think of it as a scissors enzyme) is directed to the gene that you wish to modify using a 'homing device' based on the sequence of that gene. The nuclease then snips the DNA. Three things can then happen.

1. The DNA reforms without the gene being repaired, in which case, the gene is disrupted. This might be useful in reducing food waste if that gene is involved in fruit going rotten: disruption of the gene may result in the fruit having a longer shelf life. No new genes have been added in the process.
2. The DNA reforms, with the gene being repaired using a very specific gene sequence—in which case, the gene has been edited. This may be useful in enhancing a plant's ability to fight infection. No new genes have been added in the process.
3. The DNA reforms, with the gene being repaired using a very specific gene sequence plus the addition of another new gene. This would be a sophisticated way of genetically modifying the plant.

One advantage of such innovations is that most of these new methods can be used in such a way that no new DNA is introduced into the seed sold to and grown by a farmer, although they have been modified at one point in their development. This means that, at least in principle, they would not be classified (and in Europe, therefore, labelled) as being genetically modified. This makes such modifications easier for some people to accept.

Figure A CRISPR-Cas facilitated DNA repair. Repair without a template occurs through the non-homologous end-joining (NHEJ) pathway, and this can be used to disrupt function of a gene, effectively deleting it. Repairs using a template through the homology-directed repair (HDR) pathway can enable both precise alterations and specific insertions from the template DNA sequence into the genome

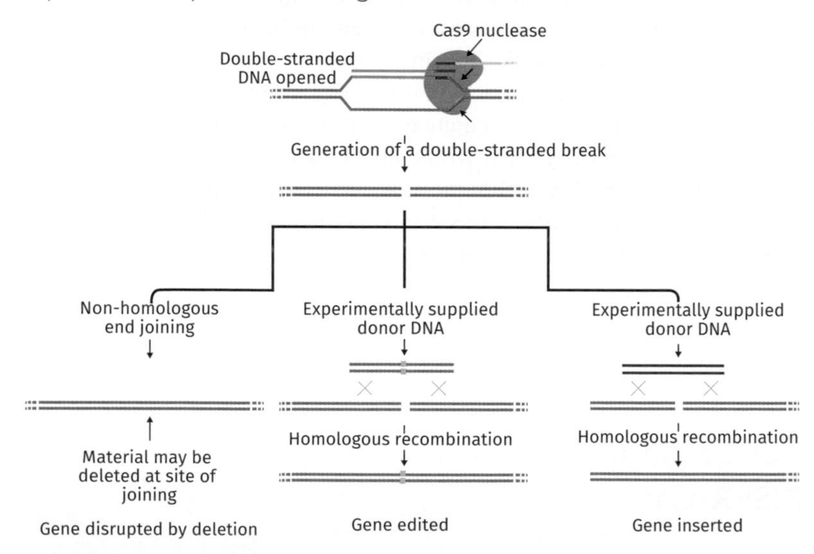

Adapted from Figure 19.35 in Craig et al, *Molecular Biology 2/e* (Oxford University Press, 2014)

Pause for thought

If a plant breeding innovation such as CRISPR/Cas9 uses 'GM technology' to introduce a desired mutation into a plant, should it be considered to be genetically modified even if the seed sold to a farmer is not genetically modified? Discuss the ethics of your opinion.

The potential impact of these new gene technologies on the future of food production cannot be over-stated. While there will be problems and huge successes along the way, these developments hold the promise of a future when we will be able to produce disease-resistant crops which can grow well and produce high yields in many different environments, helping to feed the world—and soak up a lot of carbon dioxide from the atmosphere at the same time. Look at Case study 7.2 to see two examples of these approaches in action.

Case study 7.2
Protecting potatoes

Potatoes are not just good for chips. They are of enormous importance around the world, both as a source of basic carbohydrates and as crops vital to national economies. The success of potatoes, in the past and in the future, depends on selective breeding and genetic modification.

Late blight in potatoes, caused by *Phytophthora infestans*, famously resulted in the Irish potato famines between 1845 and 1852 and still causes huge problems for potato growers. Despite new strains of potatoes bred to be resistant to *P. infestans* coming onto the market, new strains of the fungus keep arising, which are able to overcome the effects of the potato resistance genes. As a result, farmers are reliant on the use of fungicide sprays to ensure they are able to supply the market with sufficient supplies of one of the UK's staple food crops. In a bad year, with high disease pressure, farmers are forced to spray up to 17 times in one season.

One way of increasing the robustness of a potato's inherent ability to fight off infection, and thereby reduce the number of times a farmer has to spray, is to have multiple resistance genes in one potato. This means the fungus has to be able to overcome the products of more than one gene at a time, something much less likely to happen than a spontaneous ability to deal with just one issue. This is a bit like a person dealing with the dangers of walking across the road: it is relatively easy to avoid a car coming towards you; dealing with a cyclist or skateboarder on the pavement near the road is also something you can deal with; the presence of a wasp buzzing around your head is annoying but is not going to stop you. But trying to avoid a car on the road, with a skateboarder crossing your path whilst you try to avoid getting stung is not very easy at all!

New 'resistance genes' in potatoes have been found on regular occasions but getting multiple genes in one potato poses plant breeders a lot of problems. It is possible to find them in wild potato species and, in theory, it is possible to transfer the resistance from the wild to the cultivated potato. However, the resistance genes lie in close proximity on the potato genome to genes that give the resulting potatoes a bad taste. As a result, it has proved impossible to transfer the resistance genes without transferring the 'bad taste' genes.

This impasse has only been overcome recently by scientists at the Sainsbury Laboratory at the John Innes Institute in Norwich, who have cut out the resistance genes from the wild potato, cleaned up the 'bad taste' genes, and inserted the resultant clean resistance genes into cultivated potatoes. This genetic modification has been extremely successful at yielding potatoes which are highly resistant to *Phytophthora* infection, even when the disease pressure has been high, reducing the number of sprays of fungicides to two or three.

Potatoes from this programme of research are now being grown on a commercial scale in the USA, with great success at reducing losses to late blight, as you can see in Figure A. Unfortunately, at the moment such advances are not available in the UK because of a ban on the use of GM crops in Europe.

Pause for thought

Many people like seedless fruit but they also like disease-free fruit which is plentiful and cheap. Discuss the conflicts these demands raise for growers. How can new technologies help overcome these dilemmas?

Chapter summary

- Cultural control involves the management and reduction of plant diseases based on the way the plants are grown.
- Knowing how a pathogen survives and spreads within and between crops enables us to organize where and how plants are grown to minimize or prevent the spread of disease.

- There are many ways that you can manage and reduce plant disease, including chemical control. Chemical control has been used for thousands of years, but in today's world the way in which it is applied is increasingly precise, such that environmental damage is limited as much as possible.
- It can take many years of research and development to create a new chemical product which is made of active ingredients. Once developed, a product must undergo stringent health and safety assessments conducted under the European Registration Regulation, which can take up to three years to be approved.
- Just like antibiotic resistance in humans, resistance of plant pathogens is an increasing issue.
- The management of pathogen resistance is playing an increasing role in ensuring the long-term effectiveness of fungicide products. If the speed of resistance to products is faster than the development and approval of products, farmers start to lose out.
- Plant breeding has played an important role in crop production for thousands of years. Traditionally, this has involved crossing plants that have natural variation.
- Over time we have been able to become more targeted and selective of which characteristics of a plant we would like to develop or take away by mutation breeding, predicting which genes are disease-resistant, genetic modification, and gene editing.

Further reading

www.bspp.org.uk

This gives information on careers in plant pathology. If you look very closely in the careers section you can spot Kerry Maguire who wrote part of this chapter, describing her old job.

https://www.cropscience.bayer.com/en/crop-compendium

This website gives you more information on what commercial crops are grown around the world and the pests and diseases that they face.

https://croplife.org/

This site introduces real farmers from around the world and explains some of their problems growing high-quality food.

https://www.jic.ac.uk/

The John Innes website helps explain possible career routes for plant scientists interested in any aspects of plant health and disease.

https://www.rothamsted.ac.uk/

Here you can find out more about the practical aspects of being involved in agricultural science by looking at what Rothamsted Research is up to.

Discussion questions

7.1 Discuss the difference between genetic modification and genetic engineering, and think of ways in which they might benefit both the people who grow the plants and the people who eat them.

7.2 A grower wants to get the best possible yield of a crop which can be very susceptible to fungal diseases. List all of the considerations a grower would use to decide whether to use chemical control, and what product to apply to a crop.

7.3 What role might climate change play in current and future disease management?

8 BIOLOGICAL CONTROLS AND INTEGRATED CROP MANAGEMENT

Dr Julian Little

Examine any leafy plant during the warm, sunny months of summer and you will be almost certain to find animals on it. Bees and hoverflies searching for nectar, aphids tapping into the sugar-rich sap, larvae feeding continuously to fuel their rapid growth—all these are part of the natural interdependence of living things. Or are they? The caterpillar munching steadily through a healthy green leaf just might be part of an army fighting for the survival of another species—us!

Once the basic crop husbandry decisions of variety selection and cultural control have been considered, growers need other tools to prevent or treat plant diseases that may still infect their crops. Chemical sprays using synthetic fungicides or insecticides are the most well-known options to control plant diseases or the insect vectors that carry them. But there is growing interest in the use of biological control methods, using living organisms or the products of living organisms, to control plant pathogens and reduce the impact of plant diseases on crops—see Figure 8.1. This chapter will focus mainly on agricultural systems, although there may well be potential for managing plant diseases such as Chalara ash dieback and Dutch elm disease in natural ecosystems in the future.

Figure 8.1 Ladybirds have been used as biological pest controls for centuries—and they are still effective today

PHOTO FUN/Shutterstock.com

What is biological control?

Biological control uses one living organism to control the activities of another organism. In a natural ecosystem (a community of interdependent living things and their physical environment) a balance is set up. This balance is between the plants which provide the primary source of food (the producers), the animals which eat the plants (primary consumers), and animals which eat the animals which eat the plants (secondary consumers). Biological control attempts to deal with the plant pathogens or vectors of disease without destroying the ecological balance.

There are two main approaches to biological control: the use of *biocontrols* and use of *biostimulants*. These alternative methods of crop protection are very diverse and continuously evolving. As a result, many different definitions and terms are used. It can be quite confusing! So what exactly are biocontrols and biostimulants? The following are generally accepted definitions for these important new terms:

- **Biocontrols** are crop protection products which help control pests, diseases, or weeds where the active principal ingredient is a beneficial insect, a microorganism, or a compound of natural origin.
- **Biostimulants** are crop efficiency products which help promote stronger, healthier, and more robust crops where the active principal ingredient is a microorganism or a compound of natural origin.

In this primer, we will be focusing mainly on biocontrols, since these provide the greatest range of tools for direct management of diseases and pests—especially when used as part of a holistic integrated crop management (ICM) system as described in the final section of this chapter.

Why biological control?

Both biocontrols and biostimulants have been used by farmers for hundreds or even thousands of years, either knowingly or unknowingly. For example, early farmers often added manure to their soil in order to produce better crops. In doing this, they inadvertently promoted populations of beneficial bacteria and fungi with inherent biocontrol and/or biostimulant activity, helping to protect the crops against disease. Growers also recognized and actively encouraged the presence of beneficial insects such as ladybirds and lacewings, which feed on aphid pests. This both reduces direct feeding damage and also limits their impact as vectors for diseases.

Over the last century or so, we have developed many very effective chemical crop protection products, so why is there so much interest in biological control methods? In part, this interest is fuelled by the fact that we are losing many of these chemical options. In the developed world, and particularly in the UK and Europe, we have become increasingly concerned about the long-term effects of chemical crop protection products on the environment—Figure 8.2 illustrates one aspect of this. As a result, some of the world's most stringent safety checks and standards have quite rightly been developed to regulate both existing and new crop protection products. When new scientific evidence or opinion comes to light about the different compounds, they are re-scrutinized—and many uses are restricted or banned altogether. This means that there are fewer and fewer effective chemical plant protection products available for growers to use.

Figure 8.2 Neonicotinoids are a group of very effective insecticides which have had huge benefits to farmers in protecting crops against a range of insect pests which both damage crops and act as disease vectors. They have relatively low toxicity to mammals and birds which might take them in. Unfortunately it appears they may impact on the wellbeing of bees and other pollinating insects, and so in some countries their use is now restricted or banned.

borisha/Shutterstock.com

Having fewer chemical controls makes it much harder for farmers to control disease, making it difficult to achieve good yields of quality crops—and therefore difficult to make a profit. In the developing world, the cost of agrichemicals means that farmers and growers often cannot afford to use them at all. In addition, partially as a consequence of repeatedly using the dwindlingly small number of different chemical controls, we are also seeing the evolution of resistance in plant vectors and pathogens, very like the development of antibiotic resistance in the pathogens causing human disease. This is a significant concern for future disease control in these crops—or even the viability of crop production itself. As a result of all these factors, and more, it becomes harder and harder to grow the crops needed to support an ever-growing world population.

This is where biocontrols become useful—as an additional tool as part of ICM to help growers to control pests and diseases, and also to help in the fight against resistance: to help keep our chemical 'weapons' effective for when they are really needed. Biocontrols are also invaluable to organic farmers as well as those growing crops conventionally, because their natural origins mean that they are often authorized for organic use, unlike synthetic crop protection products.

Different types of biocontrols

Many different kinds of biocontrols are available for growers to use today, and these can be split into three broad categories: macrobials (beneficial insects), microbials (microorganisms), and biorationals (compounds of natural origin). These are explained in more detail in the sections that follow.

Macrobial biocontrols (beneficial animals)

The macrobial group includes a wide range of predatory or parasitic animal species which control damaging crop pests such as thrips, mites, weevils, and slugs in a variety of gruesome ways. These beneficial animals come in all sorts of shapes and sizes—from microscopic worm-like nematodes to millimetre-sized predatory mites, bugs, and parasitic wasps, and up to larger insects like the ladybirds already mentioned.

Predatory mites

One of the most common macrobials used globally, these tiny beneficial mites from the family Phytoseiidae are distantly related to spiders. Voracious predators, they are endemic natural enemies of damaging pests such as thrips, whitefly, and even other mite species such as spider mites. Particularly effective strains of these predators are reared by specialist companies and are sold to growers, who then release large numbers of them into crops to reduce the pest populations below damaging levels. This very specialist and intensive process is very expensive (tens of thousands of pounds per season per hectare). Therefore, these macrobials tend only to be used in high-value crops such as strawberries, raspberries, tomatoes, peppers, cucumbers, and ornamental plants and flowers.

Parasitic wasps

Parasitic wasp species such as *Aphelinus abdominalis*, *Aphidius colemani*, and *Aphidius ervi* are mainly used to control aphid pests such as the potato aphid, the green peach aphid, and the cotton aphid—all of which can be disease vectors in crop plants. They operate in a different way to the predatory beneficial organisms: instead of catching, killing, and eating their prey, they inject their eggs inside their hosts, where the larvae then develop, consuming the aphid from the inside out. The larvae pupate inside the dead aphid exoskeleton, emerging to start the process again, as you can see in Figure 8.3. Like the other beneficial animals, their use is largely restricted to high-value crops in protected environments.

Figure 8.3 A parasitic wasp emerges from the dead body of the aphid it has destroyed

Denis Crawford/Alamy Stock Photo

Parasitic nematodes

These microscopic nematodes are tiny worms which parasitize either slugs (in the case of *Phasmarhabditis hermaphrodita*) or a relatively wide range of soil-dwelling insect pests such as thrips, weevils, and some moth larvae (in the case of *Steinernema feltiae*). They actively seek and enter their hosts, mainly via body cavities, and then reproduce. They kill the host either because the large number of young nematodes damages the internal organs of the host or through the release of bacteria from the nematodes. The offspring then feed on the decaying corpse, seeking out further hosts to complete their life cycle.

These macrobial biocontrols are used in a wider range of crops than the previous examples, because they are both more easily applied through conventional large-scale spraying equipment and cheaper to buy. For example, slugs are major pests of many crops, including oilseed rape, potatoes, and brassica vegetables (cabbage, cauliflower, broccoli, Brussels sprouts, etc.), and farmers of many of these crops now use parasitic nematode applications to help manage this problem (Figure 8.4). This form of macrobial pest control is available to gardeners too.

Figure 8.4 Using nematodes to control slugs is a form of biocontrol available to farmers and gardeners alike

Nigel Cattlin/Alamy Stock Photo

Microbial (or microorganism) biocontrols

As the name suggests, these biocontrol agents are based on microbes: they mainly consist of bacteria or beneficial fungi, but viruses are also commonly included in this category. These work in very different ways on a wide range of targets, but are typically applied in a similar way to conventional chemical crop protection products—mostly as sprays to crop foliage or the soil. There are three main areas of microbial biocontrol of interest here: bio-insecticides, bio-fungicides, and bio-nematicides.

Bio-insecticides (or entomopathogens)

The most widely used group of these bio-insecticides are based on many different strains of the Gram-positive bacteria *Bacillus thuringiensis*—commonly known as Bt. This biocontrol has been known since the 1900s and has been actively used in agriculture to control insect pests (mainly caterpillars, but also various pest flies and beetles) for over 100 years. Bt works by producing insecticidal protein toxins inside the bacteria. Insects inadvertently eat the Bt bacteria which has been sprayed onto the plants by the farmer. The proteins are then digested in the insect gut, where the toxins paralyse the insect's digestive tract—effectively starving it to death.

The use of fungi which parasitize insects (fungal entomopathogens) in agriculture is less widespread than Bt, though it is a growing area. The most commonly used are quite widely differing strains of the species *Metarhizium anisopliae*, *Beauveria bassiana*, and *Lecanicillium muscarium*. These fungi have the potential to control a very wide range of insect pests including termites, ants, flies, beetles, bugs, caterpillars, aphids, and weevils—even mosquito and bedbug control is being investigated. Particular strains can target particular insect species, some with a wide range, some much narrower. They are most usually applied as a spray containing a suspension of fungal spores. Target insect pests then come into contact with these spores, which then germinate and penetrate the insect cuticle, growing inside until the insect dies in a matter of days. New spores are produced by the mould on the dead insect, ready to infect others.

Cydia pomonella granulosis virus (CpGV) is another widely used bio-insecticide, although it is specific to only one species of moth pest in apples and pears, the codling moth (*Cydia pomonella*). The virus has been used commercially since around the 1970s, and is sprayed onto orchards, where the codling moth comes into contact with it, becomes infected, and dies within a short period. Similarly to Bt, overuse of a specific strain of this virus led to some resistance issues, with moths appearing to be unaffected by the virus, leading to reductions in control levels. These issues have largely been overcome by a combination of other virus strains and good ICM practices (see later in this chapter), which involve a range of other control measures, including mating disruption using pheromones and chemical sprays. The use of CpGV still remains a core part of many apple and pear growers' control programmes—all designed to prevent you getting maggots in your apples!

Bio-antimicrobials

Bio-antimicrobials can be both bacterial and fungal—amazingly, many fungi can kill other fungi! Examples of bio-fungicides include the bacteria *Bacillus amyloliquefaciens*, *Bacillus subtilis*, and *Pseudomonas chlororaphis*, and the fungi *Gliocladium catenulatum*, *Ampelomyces quisqualis*, and *Coniothyrium minitans*. Many of them will also kill both fungal and bacteria plant pathogens—often producing fungicides which also have antibacterial properties.

The majority of bio-antimicrobials—both bacterial and fungal—originate from the soil, which is teeming with microscopic life: there are typically more microbes per teaspoon of soil than there are people on Earth! In this congested space, there is a never-ending battle between microbes for resources such as nutrients, water, and space. Many microbes form beneficial alliances with plants in what is called the rhizosphere (the ecosystem immediately around plant roots). These symbiotic relationships work both ways: microbes gain nutrients from the roots, and in turn they help to protect the plant from pathogens which could damage it and reduce their food source. This battleground is ideal for the development of microbes which have antifungal and antibacterial properties.

Typically, biocontrol products based on bacteria or fungi require the microbe to be in a dormant state (usually as a kind of spore) in the container to allow production, storage, and distribution to farms. Bacterial spores are much more robust and tough than fungal spores, usually needing little different in terms of storage conditions than normal chemical crop protection products. Products comprising fungal spores are typically less robust and long-lasting: they usually require refrigeration during storage, and need to be used within a short period after preparation. Sometimes, the product will also include by-products naturally made by the microbe during the process, which form an important part of the biocontrol activity.

Bacillus amyloliquefaciens and *Bacillus subtilis* are two key bacterial species with useful biocontrol properties that have been exploited by farmers. Both are rhizobacteria (root-colonizing bacteria from the rhizosphere) with plant-growth-promoting abilities–often known as plant-growth-promoting rhizobacters (PGPRs)–and they are closely related. Despite this, there are wide differences in the spectrum of biocontrol activity between them, and even between different strains of the same species.

Biocontrols based on these bacteria work in many different ways, depending on how they are applied–in the soil or on the foliage (see Table 8.1).

Because of the many ways in which bacterial-based bio-fungicides can work, they are used in a wide variety of crops and in many different ways. For example, many potato and vegetable growers are now applying these products to the soil at or near planting to colonize the new roots of their crop and help protect them against damaging soil-based pathogens such as some *Pythium*, *Fusarium*, and *Rhizoctonia* species. Growers of high-value fresh produce such as tomatoes, cucumbers, strawberries, and other soft fruit use foliar applications of specific types of these products to help protect their crops against damaging diseases like powdery mildew and *Botrytis* grey rot.

Bio-fungicides based on fungi themselves are also very important tools for many growers. These are usually parasitic or antagonistic soil-occurring fungi which attack damaging pathogenic fungi in many different ways: some are very species-specific, others are more broad-spectrum; some work

Table 8.1 Some of the ways in which *B. amyloliquefaciens* and *B. subtilis* help to control fungal and bacterial pathogens as biocontrols

In the soil	On the foliage
• *B. amyloliquefaciens* and *B. subtilis* are prolific root colonizers, often out-competing pathogenic bacteria or fungi for space on plant roots, where they consume exudates from the plants. • On the roots, they produce a sticky, slimy exudate known as a biofilm, which acts as a protective physical barrier to prevent pathogenic species getting to the roots. • The biofilm also contains bioactive compounds with antifungal and antibacterial (antibiotic) properties, which are produced by the bacilli. • The bacteria also produce plant-growth-promoting hormones which trigger effects such as increased rooting and triggers the plant's own immune system—priming it to be ready to better fight disease attacks. • Some strains also actively assist the plant to access nutrients by using enzymes to help release key elements such as phosphate from the soil—making stronger, healthier plants.	• These bacilli are soil-based organisms, so do not actively colonize foliage like they do roots. • Instead, foliar applications are mostly based around the bioactive substances which are made by the bacteria during the production process for the biocontrol product—usually a kind of fermentation. The packet of biocontrol product therefore contains both bacterial spores and the bioactive compounds. • These antifungal and antibacterial compounds are deposited on the foliage when the product is sprayed, where they act to prevent or limit fungal or bacterial pathogens from infecting the plant. • In addition, the plant-growth-promoting hormones produced by the bacteria can also act on the foliage to trigger the plant's immune system.

mainly in the soil, others can survive in the foliage—some can do both! Examples of two of the main species used for this purpose follow below.

Gliocladium catenulatum parasitizes or is antagonistic to a fairly broad range of both soil-dwelling and leaf disease-causing microorganisms such as *Fusarium*, *Pythium*, and *Botrytis*. It can be applied either to the soil or to foliage, where it can colonize and attack to prevent diseases. An interesting recent use for this bio-fungicide is for controlling *Botrytis* grey rot in strawberries, where it is delivered to the strawberry flowers by bees, which pick up the biocontrol from special dispensers in their hive. In the flowers, the *G. catenulatum* is antagonistic to *Botrytis*, preventing it from infecting the developing strawberry fruit. Most of the uses for this biocontrol agent are in protected, high-value crops such as the strawberries in Figure 8.5.

Ampelomyces quisqualis is also used as a bio-fungicide for a very specific range of diseases: this time the foliar pathogens collectively known as powdery mildews. These diseases can also infect a wide range of crops, and can have significant impacts on crop yield by reducing photosynthetic ability where the fungus infects the leaves. *A. quisqualis* applied to leaves parasitizes the powdery mildew fungus, reducing growth, or even eventually killing it altogether. This bio-fungicide doesn't survive as well outdoors, so tends to be used mainly in high-value protected crops such as cucumber and strawberries.

Bio-nematicides

Some microbes are antagonistic to or parasitic of pest nematode species—for example, the bacterium *Bacillus firmus* and the fungus *Purpureocillium lilacinum*.

These pest nematodes are different to the beneficial parasitic nematodes discussed earlier: they actually attack plant roots, causing direct damage,

Figure 8.5 Strawberries look so appealing—it's amazing to think that their perfect appearance may be the result of one fungus getting the better of another in a fruit-based fungal war—with bees delivering the weapons to the battlefield!

Anthony Short

significantly reducing crop vigour and sometimes acting as vectors for viral diseases. In many parts of the world, including the UK, they are a very significant problem for farmers.

Microbes such as *B. firmus* and *P. lilacinum* produce chemicals which are effective against these nematode pests, and this is exploited by farmers, who use them as bio-nematicide products. *B. firmus* is applied to the soil and works in a similar way to the other *Bacillus* bio-fungicide species in that it colonizes roots and produces a biofilm which provides a physical barrier against nematodes. In addition, it produces enzymes which have direct activity against nematode eggs in the soil. *P. lilacinum* is a fungus which is a parasite of nematodes, actively infecting the eggs in the soil where it is applied.

Biorationals: compounds of natural origin

Biorationals are a very diverse group of compounds that includes plant oils, natural fatty acids, and aromatic compounds such as terpenoids. Their common feature is that they are all produced naturally by microbes, plants, or animals and are harvested to be used as biocontrols. Examples include spinosad, produced by bacteria, and pyrethrins and garlic extract, which are both plant extracts.

The discovery of many of these natural compounds led to the development of synthetic chemical crop protection products—and indeed the natural world is still a source of new active substances, both for agriculture and for human medicine. However, synthetically produced versions of a molecule that can be produced naturally are specifically excluded from the biorational group.

Spinosad and the pyrethrins are probably the best known biorational control agents. Spinosad is an insecticide produced by bacteria called *Saccharopolyspora spinosa*, which were discovered in soil collected from an old sugar mill in the Virgin Islands. The active substance is effective on a wide range of insect species—both by contact and through ingestion—and is widely used across the world on a number of crops. Pyrethrins are also insecticidal, and are derived from plants of *Chrysanthemum cinerariaefolium*. The activity of this insecticide has been known for centuries: the Chinese were known to have crushed the plants and used the powder as an insecticide as early as 1000 BCE, as you saw in Figure 7.4. These compounds led to the development in the late twentieth century of synthetic pyrethroid chemistry. This group, alongside the natural pyrethrins, is still one of the most important broad-spectrum insecticide groups to this day.

Making the most of biological controls

The nature of most biological controls (with the possible exception of some of the biorationals) means that they tend to work more slowly, less consistently, and less well than their chemical crop protection alternatives. This is because, in many cases, their efficacy relies on the biocontrol altering the pathogen population in a fairly subtle way—be that predator–prey dynamics or parasite–host interactions. This, of course, doesn't mean that they don't work: they just work differently to chemicals!

In addition, biological controls themselves are influenced by things like the climatic conditions (temperature, humidity, sunlight) and are also potentially subject to direct effects from other crop management activities such as irrigation or spraying fungicides, insecticides, or herbicides.

It is therefore vitally important that farmers use biological controls in the best possible way to make the most of their benefits. This is particularly true when they are typically a relatively expensive option to use.

So, what do growers need to consider when they are using biological controls? The answer is 'lots of things', including:

- Weather conditions before, during, and after application: for example do the leaves need to be damp for the bio-fungicide to germinate and survive? Will a heavy rain shower wash them away?
- Climatic conditions after application: for example are there specific combinations of air temperature and humidity or soil temperature and soil moisture that the biocontrol requires to establish? Will there be high UV levels from sunlight that could kill the biocontrol?
- Time of day: for example should application be done at night, when temperatures are lower, humidity higher, and there is no UV light, or during the day?
- Optimal timing relative to infection levels: for example should application be made before any sign of disease or pests, in the early stages of pathogen development, or should it be targeted at a full infection as a 'fire engine' treatment?
- How should applications be made: as liquid sprays, granule applications, dips, or through irrigation systems? If sprayed, what spray coverage is required? If applied to the soil, does it need thorough mixing to ensure that spores come into contact with pathogens?
- Are there interactions (positive or negative) that need to be considered: for example, do you see a combination of biological controls with chemical applications producing a bigger effect on disease or pest control than their use alone, or do you see a lesser effect than expected—either before, during, or after the biological control is applied?
- Will the biological control need a food source to help it establish or survive when pathogen levels are low?
- For how long will the biocontrol last? For example, will it establish an ongoing colony, or will it remain only for a short period of time before normal population dynamics return the ecosystem back to an equilibrium position—such as in predator–prey cycles?

As you can see, a very complex decision-making process needs to be followed to ensure that the correct biocontrol is selected and that it is deployed in the correct way. The decision will vary depending on pest type and pressure, crop type and growth stage, temperature, humidity, day length, previous and upcoming crop sprays, and many more factors. It is not surprising, therefore, that growers often rely on highly trained specialist advisers to help them to make the correct crop protection decisions for their crops.

The Bigger picture 8.1
Exotic enemies—take care!

One thing that must always be considered with the use of live biocontrols is biosecurity. When a non-native species from one country is released to control a pest in another, it can be hugely successful. For example, the deliberate introduction of the vedalia beetle (a type of ladybird) from Australia into nineteenth-century California saved their citrus industry from destruction by the cottony cushion scale, a pest which was completely devastating the fruit trees.

However, things can go wrong when an introduced exotic organism has unintended consequences—often caused by the absence of the typical co-evolved parasites and predators of the introduced species in their new country. The best-known example of this is the introduction of the Hawaiian cane toad in the early twentieth century to Australia with the aim of controlling a sugar cane beetle pest (Figure A). Not only did the toad not manage to control the pest, which tended to live in the higher parts of the sugar cane stalks, but it also thrived in its new habitat, ousting native species and often killing them with non-native amphibian diseases as well. To make matters worse, the cane toad also produces a poison when threatened which killed many of the natural Australian predators such as goannas, snakes, and dingoes. It is difficult to see how this invasive and very damaging species can ever be removed from the countries where it was introduced.

Partly as a result of such experiences, the introduction of a new biological control agent to a country is only carried out after extensive research to make sure that it will do the job it is wanted for and nothing else. Most of the microbial biocontrols are based on species that are endemic across the world,

Figure A Biocontrols have largely been very safe—but the cane toad is a clear warning of the need for care when introducing potential control organisms into an environment

Anthony Short

and so are not usually limited in their use in this way. But many of the macro-biological control species are country-specific and so their use is strictly managed and allowed only after thorough research has been carried out. Stringent safety measures are applied when introducing totally new and alien genetic material into the environment. Up until now biological control has an excellent safety record, even when introductions of biological control organisms haven't worked as well as might have been hoped. Long may that trend continue!

Pause for thought

A fungal disease is lowering the yield of an important food crop plant. Scientists discover a type of bacterium which produces a chemical which damages and destroys the fungus which causes so much damage to the crop.

Discuss the factors that must be considered and the type of research that must be done before the bacteria can be used as a form of biological control against the fungal disease.

Integrated crop management

So far we have looked at a range of ways of controlling and preventing plant diseases, from cultural controls, with careful planning of how crops are planted and managed, through chemical controls, to the application of modern technologies such as CRISPR-Cas to develop gene edited or genetically modified plants and biological control systems. However, in the real world, control of disease is rarely solved using just a pesticide or by finding a way of growing a crop that reduces the presence of a disease. In a normal situation it is the combination of these two options in conjunction with plant breeding and possibly biological control systems that give a farmer or grower the best chance of growing a high-yielding, disease-free crop for the market place.

The best way to produce healthy crops, whether in the highly developed agricultural systems of countries such as the UK or USA, or the smaller, more localized systems in the developing world, is to use an **integrated crop management (ICM)** approach. This involves a farmer using as many of the tools in the tool box as they can. Being reliant on just one approach or 'tool' risks the pathogen that spreads the disease becoming resistant to that approach. When that happens, catastrophic crop failure can arise. But using an integrated approach, a farmer can combine different ways of reducing or preventing disease in a crop. These may include:

- selecting the best seed, whether produced by selective breeding or genetic modification;
- careful husbandry, with crop rotation built into the system
- using suitable biological control methods to minimize or destroy potential pathogens and pests;
- chemical control as and when absolutely necessary.

By using ICM, farmers can maximize their yields whilst minimizing any environmental impact, and conserving the effectiveness of both biological and chemical control systems. The effectiveness of ICM can best be seen through case studies like the one highlighted below.

Case study 8.1
Integrated crop management in action

Sclerotina stem rot is a fungal disease which affects many important crop plants, including oilseed rape and sunflowers (see Table 2.4)—both of which are important sources of plant oils used both in home cooking and in the production of many processed foods—as well as lettuces, beans, peas, and lentils. It also attacks many naturally occurring weeds, so there is a big reservoir in the environment. It is caused by the fungus *Sclerotinia*, which lives for part of its life cycle in the soil as a tough resting body called a sclerotia. Normally, this sclerotia germinates during warm and humid conditions, infecting a very wide range of crop species either through mycelia (white, thread-like strands) in the soil or through air-borne infective spores. When it infects the crop, the *Sclerotinia* fungus can invade nearly all plant tissues, producing a white cotton-wool-like mycelial growth and causing very serious crop damage (e.g. more than 50% crop reduction in oil-seed rape).

Control of this devastating pest is very important. However, rather than relying on a single management approach, an ICM system can be used to control *Sclerotinia*, minimizing the risk of infection and maximizing the benefits of bio- and chemical controls. An ICM approach includes:

- selective breeding and genetic modification to produce a range of relatively resistant plants to grow in different conditions;
- selection of the best resistant varieties of crops for the growing conditions of a particular farmer;
- rotation of the crops—because the sclerotia survive in the soil, and spores can be carried in the air from other fields, this is not a complete solution but it can help. Grasses and cereals are not susceptible to *Sclerotina* so growing them in a field which has been badly infected reduces the numbers of viable sclerotia;
- careful control of weeds, as many common weeds are affected by this stem rot and can act as a source of infection;
- using clean, disease-free seed;
- planting at the right time of year—for example in the USA, oilseed rape (known there as canola) planted in the autumn often shows less infection than seed planted in the spring;

- spraying foliar fungicides if the crop is infected. These chemical fungicides are very effective at destroying the infecting fungus but we try to use them as sparingly as possible to avoid the development of resistance in *Sclerotinia* and maintain the effectiveness of chemical control;
- biological control, including some very effective bio-fungicides based on *Coniothyrium minitans*. They are very species-specific, attacking only *Sclerotinia* species. Growers apply spores of *Coniothyrium minitans* to the soil before planting, where they come into contact with the *Sclerotinia* spores, infecting them, and preventing them from germinating to infect the crop in the first place.

Pause for thought

What are the potential benefits and drawbacks of using biological controls a) in low-value outdoor crops such as cereals and oilseed rape; and b) in protected high value crops?

In conclusion

Plants are the future of our planet. They give us food, building materials, clothing, medicines, and much more. The huge forests absorb and store carbon dioxide, and plants are the basis of almost every ecosystem on the Earth. As you have seen throughout this book, the plants we need in so many ways are threatened at every turn by diseases caused by pathogens, from viruses to fungi.

Scientists of many different types are working to find ways to prevent these diseases, and protect our crops and our environment. The approaches range from cultural changes in farming through to genome editing, and include biological and chemical controls. It is not overstating the situation to say that the future of a healthy human population depends on healthy plants both in our fields and in our natural environments. To do that we need plant scientists—over to you!

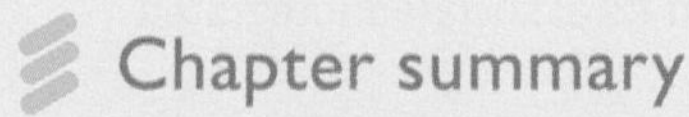

Chapter summary

- Biological control methods provide alternative ways to control and manage disease using living organisms, or products of living organisms.
- The aim of biological control is to combat and manage plant pathogens or vector-borne diseases whilst minimalizing impact on the ecological balance.
- Biocontrols are a vital tool in integrated crop management, helping growers control disease and manage disease resistance.
- They are also a key tool for organic farmers, given their natural origins.

- There are many different types of widely used biocontrols, largely falling into three categories: macrobial biocontrols, microbial biocontrols, and biorationals.
- Integrated crop management takes a holistic and realistic approach to disease management.
- For a grower, there is no one-size-fits-all solution, which means having access to as many tools in the toolbox as possible and optimizing them to manage pathogen resistance which can have devastating effects on the crop.
- An integrated approach might include seed selection, crop rotation, biocontrol, chemical control, and stewardship.
- In order to have a successful integrated crop management programme, growers must have the best toolbox available to them, which is why it is so important that scientists continue to discover new and improved ways to prevent diseases whilst protecting crops and the environment.

Further reading

- **Helyer, N., Cattlin, N.D., and Brown, K.C. 2014. *Biological Control in Plant Protection*, 2nd edition. CRC Press.**
  A recent, accessible textbook on biological control.
- **https://warwick.ac.uk/fac/sci/lifesci/wcc/research/biopesticides/amberproject/**
  The website of the AMBER project, set up at the University of Warwick to look at the practical aspects of biocontrol and how farmers and growers can use them to best effect.
- **https://leafuk.org/news-and-media/videos/integrated-pest-management-promoting-healthy-crop-growth-on-farm**
  Web resource developed by Link Environment and Farming (LEAF for short), which is an organization for farmers keen on protecting their environment, and has been promoting integrated crop management for a number of years.
- **https://ecommerce.nfuonline.com/home/ipm-plan/**
  Material produced by the National Farmers Union (NFU), who are keen to promote both integrated crop management and its associated integrated pest management.

Discussion questions

8.1 Discuss the main advantages and disadvantages of ICM.

8.2 What might be the most important problems facing plant growers in the future? Give reasons!

8.3 What issues can you predict might arise if we attempt to eliminate rather than control plant pathogens?

GLOSSARY OF TERMS

abiotic Not associatedwith/caused by living organisms. Examples of abiotic plant diseases include drought, sun-scorch, chemical damage, and root compaction.

anamorph Name given to part of the life cycle of a fungus, also referred to as the asexual form of a plant pathogen. A reproductive structure that has formed without sexual union.

antibodies Specialized proteins (immunoglobulins) produced in vertebrates as part of the immune response to invasion by pathogens. They specifically recognize a unique part of the foreign target called an antigen.

antisera Serum containing antibodies to specific antigen targets, such as virus coat protein.

aphids Sap-sucking insects, also known as greenfly. Aphids are vectors of many viruses.

bacteria Singular bacterium. Large group of unicellular, prokaryotic organisms, some of which are plant pathogens.

basidiospores Sexual spores of the basidiomycete group of fungi, formed on basidia.

basidium Plural basidia. A spore-producing structure (meaning little pedestal) unique to the fungal phylum Basidiomycotina, which supports fungal spores.

bioamplify/bioamplification The increase in concentration of a substance in tissues of an organism.

biological control The use of one living organism to control the activities of another.

biotic Associated/caused by living organisms (*cf.* abiotic).

biotroph The name given to organisms which establish a feeding mechanism with a plant and keep it alive (derived from the Greek *bios*—alive, *trophe*—feeding). For example, rust fungi keep their host alive while they feed on the plant's nutrients.

chasmothecia Singular chasmothecium, or chasmocarp. Microscopic spherical structures that house sexual spores of certain ascomycetes (in particular powdery mildew fungi).

chloroplasts The plant organelles containing chlorophyll, where photosynthesis occurs.

chlorosis Yellowing of plant leaves. This can be the result of nutrient deficiency, or from infections with a range of pathogens. Chlorosis is a general characteristic of virus infections.

circulative Virus transmission where the virus enters the body of the aphid through the midgut and circulates back to the salivary glands before further transmission can occur.

citizen science Where the general public, often working with professional scientists on specific projects, gets involved with the collection and analysis of data relating to the natural world.

classification The arrangement of organisms in taxonomic groups according to their observed similarities.

cultivars Different varieties of the same crop.

damping off Sudden death of (usually) young seedlings, often due to a combination of overly wet conditions and opportunistic organisms such as *Pythium* spp.

dicotyledonous Plants producing an embryo with two cotyledons (the first leaves to appear-from a germinating seed), includes many broad leaf crops and trees. Generally, these plants have 'radial' branching leaf veins.

DNA Deoxyribonucleic acid—a strand of genetic material made up of nucleotides which forms the basic coding for the characteristics of all living organisms. Viruses can be DNA-based, but many are RNA-based.

effector genes Parts of DNA that code for specific proteins that regulate biological activity—produced by some organisms in plant defence responses.

encapsidation The surrounding of a viral nucleic acid in a protein coat (capsid) to form a virus particle.

enzyme-linked immunosorbent assay (ELISA) A biochemical testing process which uses antisera to detect viruses in samples.

epidemiology The study of a disease—aspects such as how it spreads, what can be used to control it, and environmental conditions that affect a disease.

epidermis The outer layer of cells of a plant.

eukaryotic Member of the Eukaryotes, a group of organisms which all have cells containing membrane-bound organelles and with the genetic material contained in a membrane-bound nucleus.

facultative pathogen A pathogenic organism that can complete its life cycle without the need of a host plant.

food security Relating to the adequate supply of, and access to, safe, nutritious food to meet an individual's needs to maintain a healthy and active life.

fungicide A chemical that kills fungi with varying degrees of specificity.

fungus A diverse group of spore-producing eukaryotic organisms that lack chlorophyll and feed by decomposing and absorbing nutrients from organic materials.

genomic sequencing techniques Techniques used to determine the order of component nucleotides, or bases, in DNA or RNA.

genotype The set of genes in a cell or an individual organism or group of organisms which determine a single trait or a complex of traits.

Geographic Information System (GIS) A computer-based system that connects data with geography. It allows information to be stored, analysed, manipulated, and presented visually usually on a map.

Gram-negative Gram-negative bacteria lose crystal violet cell wall stain (and take the colour of the pink safranin counterstain) during Gram staining.

Gram-positive Gram-positive bacteria retain the crystal violet cell wall stain during Gram staining and appear purple.

haemolymph A fluid in insects and other invertebrates which is the equivalent of blood.

haustorium A specialized structure of some plant pathogens that penetrates a plant and withdraws water and nutrient from them.

herbicide A chemical that kills plants with varying degrees of specificity.

heterotrophic Organism that cannot make its own food and so has to feed on other organisms.

host plants The range of plant species that can harbour a particular pathogenic bacterium.

hydathodes Specialized pores on the leaves of higher plants that function in the exudation of water.

hydroponic A method of growing plants without using soil. Instead, plants are often grown in a soilless nutrient solution.

hypersensitive response A mechanism, used by plants, to prevent the spread of infection by microbial pathogens. It is characterized by the rapid death of cells in the local region surrounding an infection and serves to restrict the growth and spread of pathogens to other parts of the plant.

identification The use of classification criteria to distinguish certain organisms from others, to verify the authenticity or utility of a strain or a particular reaction, or to assign an existing name to an organism.

insecticide A chemical that kills insects with varying degrees of specificity.

integrated crop management (ICM) A holistic approach to agriculture which uses many different methods to maximize yield and reduce the impact of disease.

International Plant Protection Convention A multilateral international treaty which aims to produce coordinated action between countries to prevent and control the introduction of plant pests.

isometric Have a regular 3-D geometric shape.

lesion An area which is damaged through injury or disease.

metabolites Substances that are produced during metabolism or that take part in metabolism. Metabolites have various functions, including fuel, structure, signalling, stimulatory and inhibitory effects on enzymes, defense, and interactions with other organisms (e.g. pigments, odorants, and pheromones).

mildewicide A chemical which kills the pathogens that cause mildews.

monocotyledonous A classification of plants producing an embryo with a single cotyledon (the first leaves to appear from a germinating seed). Includes the cereals (maize, barley, wheat) and grasses, but also can be tree-like, such as palms and bananas. Generally, these plants have parallel leaf veins.

morphology The form, shape, or structure of an organism.

mutation Changes in the genome which may change the characteristics of the organism.

mycotoxin A toxin produced by a fungus.

National Plant Protection Organization (NPPO) This is the official government service that carries out the actions described in the International Plant Protection Convention.

necrosis Areas of dead tissue.

necrotroph A parasite that kills its host and then feeds on the dead cells. A description used for some fungi and bacteria.

nematodes Microscopic worms, also known as roundworms. Many species of these are plant parasites and can vector viruses.

next-generation sequencing Also known as high-throughput sequencing, refers to a number of different modern sequencing technologies that allow rapid sequencing of DNA from multiple organisms in parallel.

nomenclature A system of names and terms used in a particular field of study or community.

non-persistent transmission A type of virus transmission where the virus is held within the feeding mouthparts of the insect (stylet). This type of transmission can be rapid, but the virus can be quickly lost from the aphid mouthparts.

obligate pathogen/parasite A parasite that needs a host to complete its life cycle. For example, unlike many bacteria and fungi, viruses are obligate pathogens and therefore cannot be grown on artificial media.

oomycete A distinct taxonomic group of organisms within the kingdom protista that are commonly referred to as water moulds.

pathogenicity The ability of an organism to cause disease.

pathogenicity factors Molecules produced by pathogens that add to their effectiveness and enable them to colonize a niche in the host plant, evade or inhibit the host immune response, enter and exit host cells, and/or obtain nutrition from the host.

pathogenicity tests The main criteria for the identification of bacteria suspected of being the aetiological agents of a plant disease. The procedure involves inoculation of a known host plant with the bacterium, observation of the development of symptoms under disease-conducive conditions, and re-isolation of the bacterium from the diseased plant.

pathologist A scientist who studies the causes and effects of diseases.

pathovar A bacterial strain or set of strains with the same or similar characteristics, which is differentiated at infra-subspecific level from other strains of the same species or subspecies on the basis of distinctive pathogenicity to one or more plant hosts.

pesticide A chemical that kills organisms regarded as pests.

persistent transmission Virus transmission where the virus passes through the midgut of the insect into the haemolymph and to the salivary glands. There are two types of persistent transmission: 'circulative', where the virus just cycles through the insect, and 'propagative', where the virus replicates in the body of the insect as well as in the plant.

phenotype The composite of an organism's observable characteristics or traits, such as its morphology, development, biochemical or physiological characteristics, and behaviour as a result of the interaction of its genotype and the environment.

phylogenetic relatedness The relationship between an organism and other species or groups of organisms according to evolutionary similarities and diversification.

phytoalexins Substances produced by plant tissues in response to contact with a parasite that specifically inhibit growth of that parasite.

phytopathogenic bacteria Bacteria that cause disease in plants.

phytopathology The study of plant (phyto-) diseases.

phytosanitary regulation Rules made by government or other international organizations concerning plant health.

plasmids Genetic structures in a cell that can replicate independently of the chromosomes, typically a small circular DNA strand in the cytoplasm of a bacterium.

polymerase chain reaction (PCR) A molecular testing process which uses bacterial enzymes to generate copies of DNA, amplifying small samples so they can be used in DNA sequencing and profiling.

polyphasic taxonomy A consensus type of classification that takes into account all available phenotypic and genotypic data.

potyviruses A genus of viruses, many of which can cause significant damage to agricultural and horticultural crops. Most are spread by aphids in a non-persistent manner.

quorum sensing The regulation of gene expression in response to fluctuations in cell population density. Bacteria use quorum sensing to coordinate gene expression according to the density of their local population, e.g. expressing pathogenicity genes only after a threshold population of the bacteria is reached.

recombinant A type of genetic mutation where a section of one virus can join to the end of a different virus or strain, resulting in a whole new emerging strain. These changes can sometimes result in a virus gaining an evolutionary advantage.

replication The mechanism of virus reproduction; the process where a virus hijacks the functions of the cell to produce component parts of new virus structures.

reverse transcriptase An enzyme which is used by some viruses to produce complementary DNA (cDNA) from an RNA template. In virology, the same mechanism is used during PCR to enable testing of RNA-based viruses.

rhizosphere The region at the plant root–soil interface in which the chemistry and microbiology is influenced by root growth, respiration, and nutrient exchange.

RNA Ribonucleic acid—a strand of basic genetic material, like DNA, which acts as the genetic code for many viruses. RNA plays an important role in protein synthesis in cells.

seed potato certification schemes Official programme of registration and inspection to ensure that seed potatoes meet minimum standards on diseases and quality.

seed transmission Pathogen movement from a contaminated or infected seed resulting in infection and disease in the developing seedling after the seed is planted.

semi-persistent transmission Transmission where the virus is held in the foregut of the insect and the insect acts like a flying syringe.

spatial Relating to space—in this case transmission over distance, such as short-distance insect spread, or long-distance spread through plant trade.

sporangium Plural sporangia. A sporing enclosure seen in oomycetes—these can germinate directly, or act as a container for zoospores.

sputum A mix of mucus coughed up from the lungs and saliva.

stomata Openings in the epidermis of a plant that permit the passage of water vapour and absorption of carbon dioxide necessary for photosynthesis from the air, as well as the removal of excess oxygen.

stylet The mouthpart of an insect, such as an aphid, which pierces the plant and is used as a feeding tube.

symbiont An organism that lives in close and prolonged interaction with another species.

symbiotic relationship A close and long-term biological interaction between two different biological organisms, when at least one organism benefits. In a mutualistic symbiotic relationship, both organisms benefit.

systemic/systemically Affecting the whole of a plant rather than a specific part, e.g. some pathogens move throughout the plant's vascular tissues (xylem and phloem).

taxonomy The branch of science concerned with classification and naming of organisms in an ordered system that is intended to indicate natural relationships, especially evolutionary relationships.

teleomorph Name given to part of the life cycle of a fungus, also referred to as the sexual form of a plant pathogen. A reproductive structure that has formed by sexual union. Fungi are preferably named by this state, if known.

temporal Relating to time–in this case virus transmission over a time period, such as bridging between growing seasons.

thrips A type of small sap-sucking insect capable of transmitting important plant virus diseases, such as *Tomato spotted wilt virus*.

transcription The production of RNA as a template for translation by the enzyme RNA polymerase.

transient Lasting only for a short time: referring to insect populations which are not able to feed on a host but will still probe and then move on, but can transit virus during the probe activity.

translation The cellular process in which messenger RNA (mRNA) is decoded and used to produce proteins. During virus infection viruses hijack this cellular process to produce viral components for replication.

trypticase soy agar A general-purpose, non-selective media on which a wide variety of bacteria can grow. The media contains 15 g/L agar, 15 g/L casein peptone, 5 g/L sodium chloride, and 5g/L soya peptone.

tyloses Outgrowths of plant xylem cells that block the path of pathogens, preventing further plant damage.

vector Any organism, such as a sap-sucking insect, that can transmit a pathogen from an infected plant host to a healthy one.

vegetative reproduction Asexual plant propagation relating to plants which are reproduced from small, growing fragments of a parent plant such as a stem cutting, or from a tuber or bulb, rather than going through a sexual seed production cycle.

viroid The smallest known type of infectious pathogen, consisting of a short, circular, single strand of naked RNA with no protein coat.

virus A small infectious agent that replicates inside host cells.

volunteer plants A plant that has not been deliberately planted by someone, e.g. weeds in your garden, or plants growing in a field from seeds left from the previous year's crop.

yield A measure of the amount of agricultural product; in field crops this is usually expressed as the weight of produce per hectare.

zygospores Thick-walled, tough spores produced as a result of sexual reproduction in some types of fungi.

INDEX

N

O

P

Q

R

S